T0155522

Analog Circuits and Signal Processing

Series Editors:
Mohammed Ismail, Dublin, USA
Mohamad Sawan, Montreal, Canada

The Analog Circuits and Signal Processing book series, formerly known as the Kluwer International Series in Engineering and Computer Science, is a high level academic and professional series publishing research on the design and applications of analog integrated circuits and signal processing circuits and systems. Typically per year we publish between 5–15 research monographs, professional books, handbooks, edited volumes and textbooks with worldwide distribution to engineers, researchers, educators, and libraries.

The book series promotes and expedites the dissemination of new research results and tutorial views in the analog field. There is an exciting and large volume of research activity in the field worldwide. Researchers are striving to bridge the gap between classical analog work and recent advances in very large scale integration (VLSI) technologies with improved analog capabilities. Analog VLSI has been recognized as a major technology for future information processing. Analog work is showing signs of dramatic changes with emphasis on interdisciplinary research efforts combining device/circuit/technology issues. Consequently, new design concepts, strategies and design tools are being unveiled.

Topics of interest include:

Analog Interface Circuits and Systems;

Data converters;

Active-RC, switched-capacitor and continuous-time integrated filters;

Mixed analog/digital VLSI;

Simulation and modeling, mixed-mode simulation;

Analog nonlinear and computational circuits and signal processing;

Analog Artificial Neural Networks/Artificial Intelligence;

Current-mode Signal Processing; Computer-Aided Design (CAD) tools;

Analog Design in emerging technologies (Scalable CMOS, BiCMOS, GaAs, heterojunction and floating gate technologies, etc.);

Analog Design for Test;

Integrated sensors and actuators; Analog Design Automation/Knowledge-based Systems; Analog VLSI cell libraries; Analog product development; RF Front ends, Wireless communications and Microwave Circuits;

Analog behavioral modeling, Analog HDL.

More information about this series at http://www.springer.com/series/7381

Temesghen Tekeste Habte • Hani Saleh
Baker Mohammad • Mohammed Ismail

Ultra Low Power ECG Processing System for IoT Devices

 Springer

Temesghen Tekeste Habte
Khalifa University
Abu Dhabi, UAE

Baker Mohammad
Khalifa University
Abu Dhabi, UAE

Hani Saleh
Khalifa University
Abu Dhabi, UAE

Mohammed Ismail
Wayne State University
Detroit, MI, USA

ISSN 1872-082X   ISSN 2197-1854   (electronic)
Analog Circuits and Signal Processing
ISBN 978-3-030-07285-8   ISBN 978-3-319-97016-5   (eBook)
https://doi.org/10.1007/978-3-319-97016-5

This Springer imprint is published by the registered company Springer Nature Switzerland AG
The registered company address is: Gewerbestrasse 11, 6330 Cham, Switzerland

Contents

Abbreviations

ACLT	Absolute-value Curve Length Transform
AHA	American Heart Association
CAN	Cardiac Autonomic Neuropathy
CLT	Curve Length Transform
CPU	Central Processing Unit
DWT	Discrete Wavelet Transform
ECG	Electrocardiogram
HRV	Heart Rate Variability
IoT	Internet of Things
MMP	Maximum Modulus Pair
PAT	Pan And Tompkins
SoC	System on Chip
SVM	Support Vector Machine
VA	Ventricular Arrhythmia
WT	Wavelet Transform

List of Figures

List of Tables

Chapter 1
Introduction to Ultra-Low Power ECG Processor

1.1 Motivation

Health monitoring is becoming increasingly vital to the modern day community since life expectancy has risen and healthcare costs are increasing [1]. The death count due to cardiac problems is the highest when compared to other diseases [2]. Hence, there is a demand for low cost and reliable wearable biomedical sensors capable of monitoring patients and connecting them to their doctors before they reach critical situations. Wearable biomedical devices include heart rate sensors, ExG sensors, and blood glucose monitors. These portable devices are mobile and battery powered, which place a strict requirement on the power budget of the devices.

Electrocardiogram (ECG) is a bio-signal which represents electrical activity of the heart. It is widely applied in the medical field, due to its non-invasiveness and its capability of detecting cardiac diseases. ECG is normally recorded in the hospital or clinical centers in which the patient needs to stay in the hospital for hours or days. Portable or ECG telemetry devices have enabled patients to monitor their ECG, record ECG data, and transfer it to the hospital. The transferred data is processed in health centers for any abnormalities. Recent developments in Internet of Things (IoT) have enabled continuous healthcare monitors to partially process and transmit data. Developing IoT healthcare devices has multiple design challenges and trade-offs. The main constraint of wearable IoT sensor is energy dissipation due to the limited battery-energy. Efforts have been made to power IoT devices through energy harvesting sources [3], in which the system needs to operate at ultra-low power consumption. There is a trade-off between local processing and data transmission. Most of the raw data could be transmitted, whereas the local processing is limited to minimal function. In this scenario, complex computations are performed in mobile phones or PCs. Another option is to integrate ultra-low power accelerators and extract certain features. Thus, the transmitted data is minimized as it transmits only

© Springer International Publishing AG, part of Springer Nature 2019
T. Tekeste Habte et al., *Ultra Low Power ECG Processing System
for IoT Devices*, Analog Circuits and Signal Processing,
https://doi.org/10.1007/978-3-319-97016-5_1

processed data and transmits raw data during critical situations. This configuration saves the energy dissipation from the energy-hungry transmitters.

There is huge interest in the literature on ultra-low power processors or accelerators that could be integrated into IoT healthcare devices. General-purpose processors or micro-controllers dissipate more power relative to custom designed digital signal processors, which makes them unsuitable for most ultra-low power wearable devices that target long lifetime. It is desirable to design application specific chips that meet the desired performance in energy constrained configuration.

The rest of this chapter presents the design challenges and ultra-low power techniques followed by the book contribution.

1.2   Design Challenges and Ultra-Low Power Techniques

A typical wearable ECG IoT device is shown in Fig. 1.1. It consists of electrodes (to collect the analog ECG data), analog front end (to amplify and digitize the ECG signal), digital processor (to process digitized data), power management unit (to provide and control power), and wireless transmitter (to send data to remote device) [3]. For this system to have long lifetime operation, the power budget is limited to the microwatt range. It is mandatory that the power management unit controls the power sequence for each of the modules. Since the transmitter is the most power hungry modules, it should operate for short duration depending on the design requirements. Another part which consumes substantial amount of power is the digital processor, which could be a general-purpose CPU or custom designed accelerators.

In this research work, an ECG processor which consists of all the digital circuitry is designed and implemented. The digital processor which is shown in Fig. 1.1 is contrition of this research as part of an integrated IoT ECG device. In order to realize an ultra-low power ECG processor, the design challenges are summarized into two categories: architectural choices and circuit design techniques.

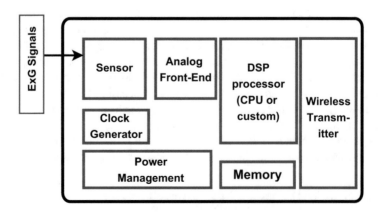

Fig. 1.1 ECG node for IoT healthcare system

The architectural choice is determined by the required application and the available power budget. The digital processor is designed in order to minimize the power consumption and energy dissipation. There are multiple ways of implementing a digital system either by utilizing a central processing unit (CPU) or by developing fully custom design. Hardware accelerators, specific to an algorithm or application, are more computationally and energy efficient. An algorithm implemented in an ASIC is more energy efficient than an algorithm implemented in a general-purpose processor or even low-power micro-controllers. Integrating a CPU along with custom accelerators provides a balance between energy efficiency and flexibility. Efforts have been made to minimize the energy consumption both at the architectural level and at the device level while attaining the desired application performance [4, 5]. ECG sensing devices have been researched and ECG signal contains wave components such as QRS complex, T-wave, and P-wave. The QRS complex is the dominant part with highest slope and magnitude. The ECG wave components (QRS complex, T-wave, and P-wave) along with their peaks and edges are termed as ECG features. Several ultra-low power ECG processing architectures were reported for application in QRS detection [6–11], full ECG feature extraction [12, 13], and ECG classification [8, 12]. Some of these architectures comprise a CPU [7] that consumes relatively higher power. Others are more energy efficient and have fully custom designs. Most of the reported QRS detection architectures are based on wavelet transform [6, 8–10]. Wavelet transform is robust in detecting QRS complex, however, its implementation requires multiscale decomposition using a cascade of filters. Others have reported QRS detection architectures based on an algorithm known as Pan and Tompkins (PAT) [11, 12]. The PAT algorithm is based on filtering, differentiation, squaring, and moving integral. All of these operations require hardware-intensive computations. Wavelet transform using multiscale decomposition has been applied also for full ECG feature extraction [13]. Time domain signal analysis is applied [12] for full feature extraction. The challenge in QRS detection or ECG feature extraction is to attain the desired performance at the lowest possible power consumption. That is why appropriate architectural choice is necessary.

Circuit design techniques that minimize the energy consumption are widely explored and implemented in current state-of-the-art portable medical devices. These techniques include voltage scaling, voltage islands, power gating, clock gating, and the use of deeply scaled nodes.

The total power consumption in an ultra-low power digital system can be represented by Eq. (1.1). According to this expression, the factors that affect the total power consumption are:

$$P_{Total} = \sum (\alpha_i \times C_i) \times f \times V^2 + \beta \times I_{leakage} \times V \qquad (1.1)$$

- C_i: load capacitance
- V: supply voltage

- f: operating frequency
- α_i: switching activity factor
- β: leakage factor

Each of the above factors could be reduced to minimize the total power consumption. Firstly, the load capacitance depends on the technology, and it could not be altered as long as the same standard cell library is used. Secondly, scaling the supply voltage will reduce the power. The dynamic power is minimized quadratically and the leakage power is minimized exponentially with the supply voltage. It is an effective technique; although, there are constraints to limit the supply voltage. These constraints are the minimum operating voltage of SRAM and the timing requirement of the design. Thirdly, the operating frequency must be optimized and is determined by the architecture of the system. Thus, choosing an optimum frequency would enable further reduction in power consumption. Reducing the operating frequency will reduce the dynamic power as represented in Eq. (1.1). Fourthly, the switching activity factor depends on the architecture, and it could be altered by applying clock gating. Certain blocks could be clock gated whenever they are not used. Hence, applying a combination of the above techniques would enable the system to achieve the minimum average power consumption and henceforth the lowest energy consumption. Lastly, the leakage factor could be reduced by utilizing low-leakage standard cell library. Certain parts of the design could be power gated in case they are not used at certain time slots, as power gating reduces the leakage.

The digital processor could be powered from energy harvesting devices or battery. In recent developments, energy harvesting circuits were capable of powering full SoC that includes analog front end, digital processor, and a transmitter [14]. It is necessary that the power consumption remains within the power limit of the energy harvester. If the system is operating at its lowest energy dissipation, it would have a long lifetime. Hence, the proposed approaches would investigate achieving lowest possible power consumption and minimum energy dissipation.

1.3 Book Contribution and Organization

The main objective of this research work was to design an ultra-low power ECG processor for IoT healthcare devices. Ultra-low power operation was achieved through novel architectural choices and ultra-low power circuit design techniques. In this book, two chips were fabricated and tested. The fabricated chips process ECG signals in order to extract features and classify ECG beats for cardiac autonomic neuropathy (CAN) severity. Moreover, an improved architecture for ventricular arrhythmia (VA) prediction is proposed. The designed chips were purely digital circuits operating in the nano-watt range. The main system components are shown in Fig. 1.2 and include ECG processing accelerators to extract and classify ECG beats, custom designed control FSM that controls the overall chip functions, memory to store temporary data, and digital interface for external communication. Full-custom design techniques resulted in a much lower power dissipation than did the incorporation of a general-purpose processing unit.

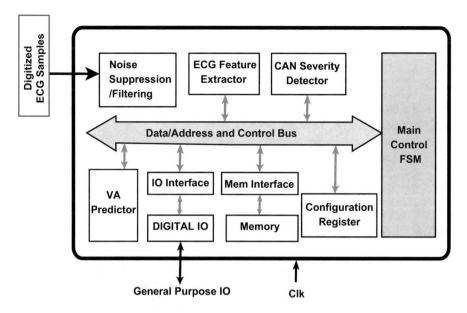

Fig. 1.2 Integrated biomedical processing platform

The contributions of the book work could be summarized as follows.

1. Combined CLT and DWT-based ECG processor (Chap. 3)

   - A combination of CLT and DWT architectures for full ECG feature extraction, where each technique is chosen for its advantages in reducing hardware resources without affecting the desired performance.
   - Developed a pipelined architecture for CLT which reduces the required resource by $32\times$ when compared to conventional straightforward implementation of CLT.
   - Designed and implemented ultra-low power techniques such as clock gating and voltage scaling. Through such techniques, the fabricated chip consumed only 642 nW at 0.6 V and at a frequency of 7.5 kHz.

2. Ultra-low power QRS detection architecture (Chap. 4)

   - An ultra-low power QRS detection architecture was designed that is capable of detecting QRS at sensitivity and predictivity of 99.37% and 99.38%, respectively.
   - Absolute Value Curve Length Transform was proposed, where the required hardware resources are minimized.
   - Synthesized QRS detection system consumed 6.5 nW when operated at 1 V and at 250 Hz.
   - A lossless compression architecture that enabled reduced transmitted data for IoT transmitters was implemented.

3. Ultra-low power CAN detection and VA prediction architecture (Chap. 5)

- On-chip full ECG feature extraction that utilizes ACLT for QRS detection and low-pass differentiation for other ECG features was implemented.
- Real-time classification of CAN severity is enabled through on-chip classification of CAN severity.
- Fabricated chip that extracts full ECG features and detects CAN severity that consumed only 75 nW at 0.6 V.
- Improved hardware architecture for VA prediction is presented, which achieves a reduction in the required area by 16.0% and in power consumption by 62.2%.

1.3.1 Book Organization

The remainder of this book is organized as follows. Chapter 2 discusses background about ECG processing algorithms and biomedical sensing platforms. Chapter 2 presents existing algorithms, their drawbacks, and analysis. Chapter 3 presents an ECG feature extraction SoC based on combined curve length transform and discrete wavelet transform. Chapter 3 discusses the merits of each of these transforms in ECG processing. Moreover, it presents ultra-low power techniques along with the measured results. In Chap. 4, an energy efficient QRS detection architecture along with a compression architecture is proposed. Chapter 5 discusses an ECG processor aimed at CAN detection. The SoC performs full ECG extraction and on-chip CAN detection. Measured results are presented along with comparison with literature. Chapter 6 concludes the book.

Chapter 2
IoT for Healthcare

2.1 Introduction

The Internet of Things (IoT) represents a set of interconnected smart objects and people at any time and at any place. The IoT incorporates wide spectrum that can impact businesses, healthcare, social and political aspects. It is a platform that extends from sensors, local processors, wireless transmitters, and central management stations [15].

Figure 2.1 shows the trends for IoT healthcare devices. It incorporates wide sectors that involve many individuals. The main feature of IoT healthcare platform is the communication between a wearable sensor and central system where doctors could easily assess patients. In terms of health conditions, it includes early diagnostics, emergency situation, and chronic diseases. Such connected platform also utilizes the existing voice and data communications infrastructures.

2.2 IoT Healthcare Applications

The application of IoT-based healthcare covers wide areas in the healthcare sectors. It extends from individual applications and in healthcare centers. It includes care for children, youths, elderly along with wide diversity of patients through organized system. This section describes the applications for IoT healthcare. The applications are directly related to the end-users and patients. Current connected wearable healthcare devices are a good example for IoT healthcare devices. The next subsections describe numerous IoT healthcare applications.

© Springer International Publishing AG, part of Springer Nature 2019   7
T. Tekeste Habte et al., *Ultra Low Power ECG Processing System for IoT Devices*, Analog Circuits and Signal Processing, https://doi.org/10.1007/978-3-319-97016-5_2

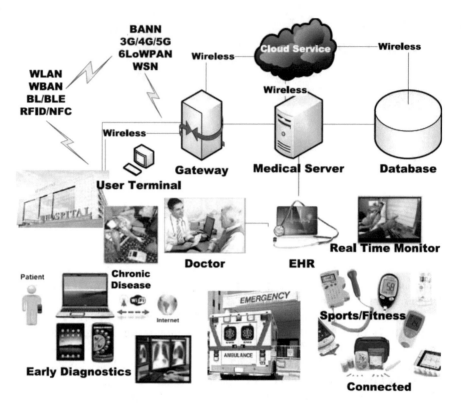

Fig. 2.1 IoT healthcare trends

2.2.1 Glucose Level Sensing

Diabetic patients suffer from uncontrolled blood glucose level during the daily lives. It necessitates the need for periodic monitoring of their blood glucose level, in order to plan their daily meals, medicines, and activities. Real-time glucose level sensing in IoT platform is reported in [16]. The reported technique demonstrates how sensors from patients are connected to respective healthcare centers through IPv6 connectivity.

2.2.2 Electrocardiogram Monitoring

The electrocardiogram (ECG) records the electrical activity of the heart which represents the full cardiac cycle. The ECG is the best known technique to monitor and diagnose the function of the heart. ECG measurements include determining the beat rate and other patterns of the cardiac cycle which is displayed in ECG waveforms. ECG techniques have been applied in prediction [12] and detection [17] of arrhythmia such as ventricular tachycardia, bradycardia, arterial fibrillation, and myocardial infarction. An IoT-based ECG monitoring platform is reported in liu2012internet and the system comprises wearable wireless ECG sensor and wireless receiving processor. It does real-time cardiac function detection and abnormality classification.

2.2.3   Blood Pressure Monitoring

IoT is also applied in the blood pressure monitoring as described in [18]. In this paper, the combination of a touch blood pressure (BP) meter and a near-field-communication for a BP monitoring systems, where also a mobile phone is utilized.

2.2.4   Oxygen Saturation Monitoring

Blood oxygen saturation is monitored using pulse oximetry technique in portable healthcare devices. Hence, it is beneficial to integrate pulse oximetry into the IoT platform. The potential for implementing an IoT-based pulse oximetry is illustrated in [19]. An IoT-centered low-power and economic low-cost blood oxygen saturation meter for remote patient monitoring is proposed in [20]. The proposed system provides continuous measurements by operating from battery.

2.3   IoT Healthcare Technologies

IoT-based solution has been enabled through wide range of technologies. Advanced electronic solutions have profound effect on the fast growth of IoT solutions. IoT healthcare solution is supported through many technologies. This section discusses the core technologies that have the capability to enhance IoT-based healthcare services.

2.3.1   Ultra-Low Power Sensing

Ultra-low power sensors that include sensors, analog front end, and digitization form a key part for signal acquisition in IoT healthcare devices. These components which have ultra-low power dissipation, could be integrated in battery powered systems and attain long battery lifetime.

2.3.2   IoT Processors

The processing part of an IoT device could perform from simple to complicated tasks based on the desired application. Some of the main tasks are signal acquisition, local processing, data transmission, security, and encryption. Local processors could support an operating system, firmware, and device management. The main constraint of local processors is power consumption. There is a trade-off between power dissipation, supported features, hardware costs, and software costs. Hence, deep analysis of the system is necessary to select appropriate selection of the processor.

2.3.3   Cloud Computing

Cloud computing facilitates resource sharing among various IoT healthcare technologies and devices. Its capability enables service upon request for specific resources through the network.

2.3.4   Grid Computing

Since medical sensor nodes have limited processing capabilities, complicated computational processing is supported by grid computing. Grid computing forms the backbone of cloud computing.

2.3.5   Big Data

Large amount of data is normally collected by wide range of medical sensors. Systematic handling and processing of big data improves the efficiency of the health diagnosis and monitoring. Moreover, disease classification is performed through big-data analysis.

2.3.6   Communication Networks

IoT-based healthcare networks share the existing communication networks that could be short range (WLAN, 6LoWPANs, WBANs, WBANs, and WSNs) or long-range communications (cellular networks). Moreover, other wireless communication networks such as bluetooth, near-field communication, and RFID communication technologies are powerful instruments in achieving ultra-low power IoT healthcare sensors.

2.3.7   Wearable

Wearable technology has enabled continuous monitoring and active engagement of patients. Wearable devices bridge the communication between the patient and the doctors. Local processing in wearable sensors has also a life saving benefits through alerting patients on time before they reach critical situations.

2.4   IoT Challenges in Healthcare

IoT-based healthcare services have several challenges that arise from the sensors, communication networks, and central servers. Here, we will describe some of the main challenges of IoT healthcare devices.

2.4.1   IoT Healthcare Security

Since the IoT connected devices are increasing day by day, security is a major issue that we have to develop. The IoT is growing rapidly and it is expected widespread adoption of the IoT healthcare systems. Healthcare devices and applications are expected to deal with vital private information such as personal healthcare data. In addition, such smart devices may be connected to global information networks for their access anytime, anywhere. The protection of captured health data from various sensors and devices from illicit access is crucial. Therefore, stringent policies and technical security measures should be introduced to share health data with authorized users, organizations, and applications. This tradition of confidentiality is the most essential thing, which personal data must be obtained for a specified purpose, and must not be disclosed to any third party except in a manner compatible with that purpose. A robust system security must be introduced to prevent an attack, vulnerability, or data loss.

2.4.2   Energy Consumption of IoT Healthcare Devices

There are many devices in IoT healthcare scenarios, and such devices tend to be heterogeneous in terms of their sleep, deep-sleep, receive, transmit, and composite states, among others. In addition, in terms of service availability, each communications layer faces an additional challenge in terms of power requirements. Regardless of the type of connection behind an IoT product, minimizing power use can be a challenge. However, it is also critical to attain low energy and low costs. Radios are a key component of the energy budget for IoT products. Instead of WiFi, cellular or Bluetooth, combined with smart design, and component selection, will help developers create low-energy systems.

2.4.3 Communication Network

Various networks ranging from networks for short-range communications to long-range communications are part of the physical infrastructure of the IoT-based healthcare network. In addition, the employment of ultra-wideband (UWB), BLE, NFC, and RFID technologies can help achieve low-power medical sensor devices as well as communication protocols.

2.4.4 Data Storage and Continuous Monitoring

There is a trade-off between local processing and transmission in continuous monitoring. Moreover there is a trade-off between local data storage and central data storage. Such situations are decided based on the application and the architectural choices.

Chapter 3
Background on ECG Processing

3.1 Introduction

Wearable healthcare devices play a crucial role in our daily lives starting from monitoring our daily well-being to performing safe exercise. Health issues prediction systems would reduce the risk of sudden death through early alarms before catastrophic health situations. Among all global deaths, cardiovascular disease and sudden cardiac arrest contribute the largest percentage of deaths (31% in the year 2012 [21]). Continuous monitoring of heart activity using ECG and local processing in wearable devices would enable preventive measures for high risk individuals.

This chapter describes the state-of-the-art biomedical systems, biomedical feature extraction algorithms, and digital circuit architectures. Generally, any biomedical system consists of an analog front end, signal conditioning, signal processing, and transmission blocks as shown in Fig. 3.1. Each of these blocks has its own design requirements. All biomedical systems require an analog front end with high common mode rejection ratio in order to suppress common mode noise that is present on the output of bio-signal sensors. The signal conditioning block depends on the nature of the biomedical signal and necessary filtering. Once a clean signal is available, robust algorithm could be constructed to extract clinical information in the digital signal processing. The transmission block is utilized to transmit raw data to a central unit or processed data for display.

3.2 ECG Basics

The ECG signal represents the electrical activity of the heart that can be captured by non-invasive technique. It is one of the most prevalent and commonly used signals in the medical field. Figure 3.2 shows the heart which is the source of ECG signal and

© Springer International Publishing AG, part of Springer Nature 2019
T. Tekeste Habte et al., *Ultra Low Power ECG Processing System
for IoT Devices*, Analog Circuits and Signal Processing,
https://doi.org/10.1007/978-3-319-97016-5_3

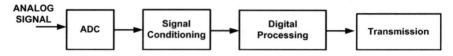

Fig. 3.1 ECG system

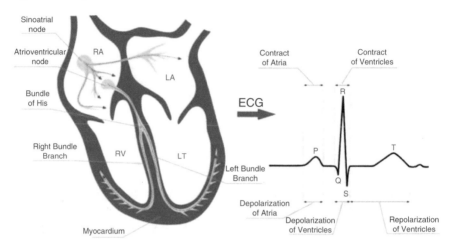

Fig. 3.2 Basic ECG signal

a typical ECG waveform. ECG is characterized by distinct morphologies, namely P-wave, QRS complex, T-wave, and U-wave. The U-wave is not commonly visible and most of time is not clinically applicable. These ECG wave components such as P-wave, QRS complex, and T-wave along with their peaks and edges are termed as ECG features. All of these components are linked to different activities in the heart cardiac cycle. The QRS complex, which is a principal component in the cardiac cycle, is used as a reference and represents the depolarization of ventricles in the heart. Its amplitude rises to 1 or 2 mV above or below the isoelectric line for normal beats and can go several times larger for abnormal beats. The time required for the ventricles to depolarize defines the QRS width or interval where it typically lasts between 80 and 120 ms [22]. Other intervals such as PR interval (from P_{onset} to Q_{onset}) and QT interval (from Q_{onset} to T_{offset}) range from 120 to 200 ms and 350 to 430 ms, respectively.

In addition to the time domain nature of an ECG signal it is necessary to investigate the frequency spectrum of an ECG signal. Figure 3.3 shows the spectrum of a typical normal ECG signal. The QRS complex has frequency components in the range 5–15 Hz. The P and T waves are concentrated in the lower frequency range 3–5 Hz. Based on their respective frequency range appropriate extraction techniques could be developed for each of the ECG waveform features.

Due to the ease of use of ECG and non-invasiveness of ECG detection, it is not only used as a prime tool to monitor the heart but also to diagnose cardiac arrhythmia

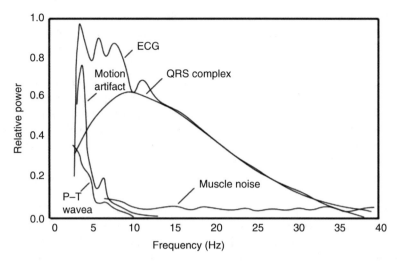

Fig. 3.3 ECG spectrum

by extracting information about intervals, amplitudes, and wave morphologies such as P, QRS, and T waves [23]. Features extracted from the ECG signal play an essential role in diagnosing many cardiac diseases. Hence, the development of real-time and accurate delineation methods is crucial especially for abnormal ECG signals.

3.3   ECG Feature Extraction Techniques

In order to develop a robust signal processing platform, the nature of ECG signal in time and frequency domain has to be taken into account. Since the QRS complex forms relatively higher amplitude and higher slope, its detection is key to automatic techniques. Various signal processing techniques for QRS detection have been proposed in the literature. Time domain thresholding along with filtering (first derivative, second derivative, both derivatives, matched filter, etc.) were some of the earliest techniques and are suitable for real-time implementation [24–26]. Other methods that provide enhanced accuracy were based on the spectral analysis of the ECG signal. In [27–30], wavelet transform was presented as a tool to analyze ECG signals. As a part of the spectral analysis techniques, discrete Fourier transform was reported in the literature to detect the QRS complex [31]. Empirical mode decomposition and Hilbert transform were introduced to improve the analysis of the QRS detection of nonlinear and non-stationary ECG signals [32, 33]. Moreover, principal component analysis (PCA) that linearly transforms the ECG data into new coordinate system was proposed in [34]. QRS detection techniques could also be based on the concept of machine learning, classification, and pattern recognition, mostly applicable when the QRS complex is used in the diagnosis of cardiac arrhythmia. Such techniques include fuzzy logic [35], artificial neural network [36], neuro-fuzzy networks [37], support vector machine [38], or combinations [39, 40].

Delineation, which is the stage of determining the fiducial points and the limits of the ECG waves, is very essential to extract the different ECG parameters such as ST interval and QT interval. The localization of wave peaks is easier than the onsets and offsets, as the signal to noise ratio is higher and becomes lower at the wave boundaries where the noise level dominates the signal, which in turn leads to a complex delineation process. Usually, ECG wave delineation is done after detecting the QRS complex where a set of search windows are defined to locate T and P waves. The search window enhances the characteristics of the targeted waves using different approaches proposed in the literature. In [41], a delineation technique based on time curve derivative of a digital signal is proposed. Adaptive filters in many different forms were also used in an ECG delineation process [42, 43]. Time domain morphology and gradient [44], hidden Markov models[45], and Bayesian approach and Gibbs sampler [46] were other methods that offer a wide range of complexity, flexibility, accuracy, and hardware implementation cost.

3.3.1   Methods Based on Discrete Wavelet Transform

Discrete wavelet transform (DWT) provides both time and frequency information without resolving all frequencies equally. At high frequencies, DWT provides good time resolution and poor frequency resolution while it does the opposite at low frequencies. Thus, it is a useful tool when the signal has high frequency components for short duration and low-frequency components for long as in ECG signal. DWT was widely used and became an important tool to analyze the ECG signal and delineate its fiducial points [27, 44].

Wavelet transform (WT) provides a decomposition of the signal over a set of basic functions, obtained by dilation and translation of a mother wavelet by a scale factor α and a translation parameter β. The WT of a signal $x(t)$ is defined as [27]:

$$W_a x(b) = \frac{1}{\sqrt{\alpha}} \int x(t) \ \phi\left(\frac{t-\beta}{\alpha}\right) \ dt \qquad (3.1)$$

WT can be evaluated with discrete mathematical operations, which leads to the formulation of discrete wavelet transform (DWT). The DWT is implemented as an octave filter bank by cascading low-pass and high-pass filters. To keep the temporal resolution at different scales, a technique named *Algorithme Trous* [47] was implemented as shown in Fig. 3.4, where $h[n]$ and $g[n]$ are given in Eqs. (3.2) and (3.3), respectively. A mother wavelet based on quadratic spline wavelet is selected due to its ease of implementation and accuracy for analyzing the ECG signals. The Fourier transform of the selected mother wavelet is given in Eq. (3.4) [27].

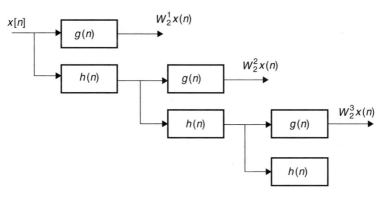

Fig. 3.4 Two filter-bank implementations of DWT (Trous algorithm)

$$[H]h[n] = \frac{1}{2} \ [\delta[n+2] + \delta[n+1] + \delta[n] + \delta[n-1]] \tag{3.2}$$

$$g[n] = 2 \ [\delta[n+1] - \delta[n]] \tag{3.3}$$

$$[H]\Phi(\Omega) = j \ \Omega \ \left(\frac{\sin(\frac{\Omega}{4})}{\frac{\Omega}{4}}\right)^4 \tag{3.4}$$

3.3.1.1   Detection and Delineation of ECG Signal Based on DWT

The ECG signal is coupled with different forms of noise such as baseline drift, sudden body movement, and power-line interference. Figure 3.5 illustrates the effect of 50 Hz power-line interference and baseline wander due to body movement. Unlike other ECG detection and delineation methods, DWT suppresses the noise in a single step without the need for pre-filtering. In this subsection, DWT based full ECG feature extraction will be presented as reported in [27]. The analysis will try to review the various aspects of ECG feature delineation. The different ECG components are visible at different DWT scales and the zero crossings of maximum modulus pair (MMP) across these scales correspond to the fiducial points.

3.3.1.2   QRS Complex Detection and Delineation

Multiscale DWT was applied for QRS detection as reported in [27]. A window of 4 s is designed to search for the QRS complex; whenever a QRS complex is detected, an eye blocking window of 200 ms is utilized before searching for the next peak. The significant slopes of the QRS complex are associated with the maximum

Fig. 3.5 Effect of artifacts on ECG. (**a**) Clean ECG. (**b**) 50 Hz corrupted ECG. (**c**) Baseline-wander ECG

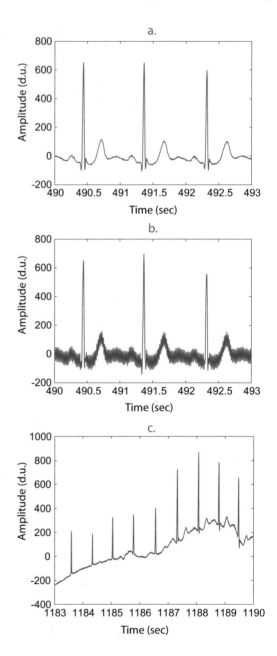

of MMP in $\mid W_2^3 x[n] \mid$. The zero crossing of MMP corresponds to an R peak. Q onset and Q offset are at the edge of these slopes before and after the R peak, respectively. A pair of maximum modulus lines at scale 2^2 before and after R peak represents Q wave and S wave, respectively. An MMP is classified to correspond

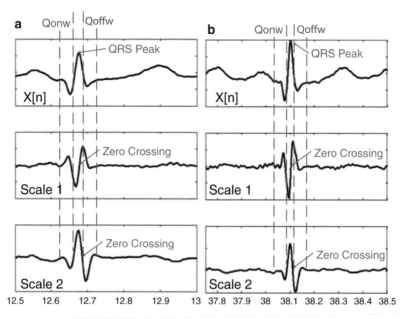

Fig. 3.6 Examples of QRS detection and delineation. The first row represents the raw ECG signal. The DWT of $X[n]$ across scales 2^1 and 2^2 are shown in rows 2 and 3, respectively

to R peak, Q onset, or Q offset based on thresholds, and the threshold values are given in Eqs. (3.5)–(3.7). Whenever ϑ_R changes, ϑ_{Qon} and ϑ_{Qon} are consequently updated in each heartbeat. A temporal window is defined before and after the R peak and the thresholds are applied to search for Q onset and Q offset. In Fig. 3.6, two examples of QRS detection based on DWT across the first and second scales are illustrated.

$$[H]\vartheta_R = 1.5 \text{ RMS } (win_{RR} \mid W_2^3 x[n] \mid) \tag{3.5}$$

$$\vartheta_{Qon} = 0.1 \ \vartheta_R \tag{3.6}$$

$$\vartheta_{Qoff} = 0.1 \ \vartheta_R \tag{3.7}$$

3.3.1.3   T and P Wave Delineation

The T wave is the representation of repolarization of ventricles whereby the myocardium is prepared for the next cycle of the ECG. The P wave is the representation of repolarization of atrial. In the automatic delineation of the ECG, it involves locating peaks, onsets, and offsets. Among all ECG features the most challenging is the detection of T wave ends. This is mainly due to the slow transition of the signal near the end of T wave. In addition, the T waves have oscillatory patterns that vary from one individual to another which make the delineation process even more challenging. After the QRS detection, two search windows at scale 2^4 are defined depending on the location of the QRS complex and the previously computed RR interval. Different size of windows is identified to look for T and P wave separately. A T wave is located in window win_T if MMP exists in DWT scale 4 ($| W_2^4 x[n] |$) and the local maxima exceeds a threshold ϑ_T. Similarly, a P wave is identified within the window win_P if MMP exists in $| W_2^4 x[n] |$ and the maxima points exceed a threshold ϑ_P. T and P wave threshold levels are given in Eqs. (3.8) and (3.9), respectively. The wave boundaries are identified by looking at the positive maxima and negative minima of their respective MMP. The zero crossing of their MMPs is mapped to their peaks. The delay due to higher scales is also taken into account when mapping to the original ECG signal. Examples of T and P wave delineation of different morphologies are illustrated in Fig. 3.7.

$$\vartheta_T = \text{RMS} \ (win_T \ | \ W_2^4 x[n] \ |) \tag{3.8}$$

$$\vartheta_P = \text{RMS} \ (win_P \ | \ W_2^4 x[n] \ |) \tag{3.9}$$

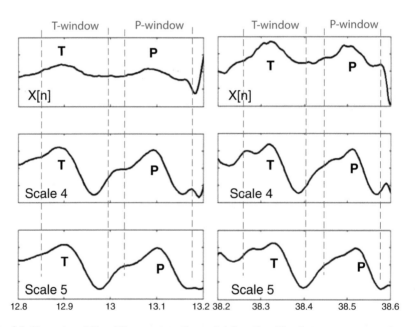

Fig. 3.7 Examples of T and P wave detection and delineation. The first raw represents the raw ECG signal. The DWT of $X[n]$ across scales 2^4 and 2^5 are shown in rows 2 and 3, respectively

3.4 ECG Classifiers

The ECG features are utilized to define parameters for further classification in order to detect or predict cardiac disease. Out of these features, many intervals such as the QT interval, PR interval, and RR interval could be defined based on the application. Some classifiers are used for detecting arrhythmia that make use of multiple parameters extracted from the ECG waveform [48]. The reported system consists of a weak linear classifier and a strong support vector machine classifier. Low power is achieved using weak linear classifier and relatively at a higher power consumption the accuracy is enhanced using the integrated support vector machine.

Others have reported cardiac autonomic neuropathy (CAN) detection system that makes use of ECG features. CAN is a cardiac disease related to diabetic patients. The CAN severity detector is described in [49]. Its main objective is to determine CAN based on heart rate variability and the QT variability index (QTVI), which will be discussed in detail in Section 6.2.3. Figure 3.8 demonstrates how the QTVI could be utilized in CAN detection and QTVI is evaluated as in Eq. (3.10).

$$QTVI = \log_{10}\left[\frac{\frac{QTv}{QTm^2}}{\frac{HRv}{HRm^2}}\right] \tag{3.10}$$

3.5 Review of Biomedical SoCs

A brief survey of the most relevant current state-of-the-art biomedical SoCs along with their architectures, complexity, performance, and power reduction techniques will be presented.

Figure 3.9 shows an architecture of a biomedical sensing platform [4] that could be used for ECG or EEG. The platform contains a 16-bit CPU with extended logic to support software debugging as well as a direct memory access (DMA) block. Moreover, accelerators that perform DSP algorithms, namely FFT, median filter, FIR filter, and CORDIC are incorporated in the system. The system achieves $10.2\times$ and $11.5\times$ energy reduction when running EEG and EKG applications, respectively, compared to CPU-only implementation. Implemented accelerators along with voltage scaling scheme and architectural optimizations contribute to the energy efficiency and substantial power requirement reduction of the system. Though the system contains multiple accelerators, it contains a power hungry CPU. The power consumption from the CPU is dominant and the CPU is always ON, which has negative impact on the energy dissipation. Replacing the CPU with custom application specific block would be more power and energy efficient.

Zhang et al. [14] developed a body-area-sensor node capable of operating at a total chip power of $19\,\mu W$. The system is powered by a thermal energy harvester

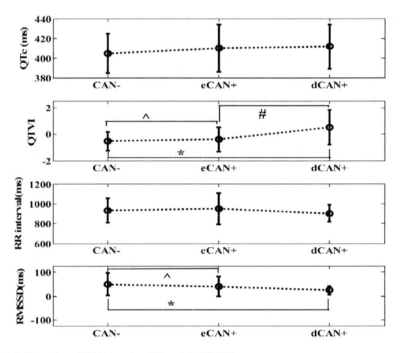

Fig. 3.8 Detection of CAN based on RR and QTVI [49]

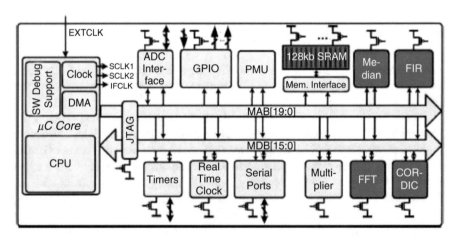

Fig. 3.9 Architecture of a biomedical signal processing platform [4]

from the human body heat as depicted in Fig. 3.10. Power reduction is achieved by employing digital power management unit and hardware accelerators such as programmable FIR, dedicated accelerator for ECG heart rate (RR) extraction, atrial fibrillation (AFib) detection, and EMG band energy calculation. A digital section provides mode control and power management enabling dynamic voltage scaling.

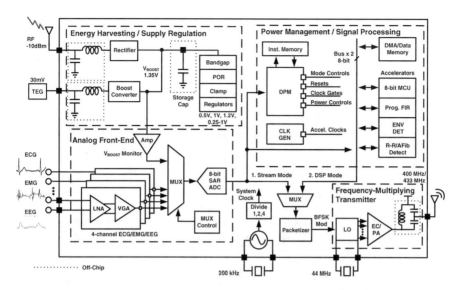

Fig. 3.10 Architecture of a biomedical signal processing platform [14]

Moreover power and clock gating techniques are implemented in order to minimize the power consumption. The system does many functions and can sense multiple signals such as ECG, EMG, and EEG. However, it still contains an always-ON micro-controller unit, which affects the energy dissipation negatively. Fully custom processing of the individual signals would be more power efficient.

Recent work in an integrated biomedical system for the Internet of Things (IoT) was reported in [3]. The system comprises photovoltaic and thermoelectric generator, analog front end, digital processor, and wireless interface as revealed in Fig. 3.11. Its key feature is that it demonstrates the highest level of integration and highest energy harvesting efficiency. It incorporates switchable power domain for maintaining ultra-low power operation.

Some biomedical chips are meant for implantable applications. The work in [50] presents implantable ECG monitoring system capable of running an ECG RR interval extraction at a power consumption of only 64 nW. The chip is powered by a thin film Li battery and contains ECG electrodes, analog circuits, digital processor, and wireless circuits as shown in Fig. 3.12.

3.5.1 Ultra-Low Power Digital Circuit Design Techniques

The power consumption in digital circuits is the sum of leakage P_{leak} and dynamic power P_{dyn} given by Eqs. (3.11)–(3.13) [51]. The P_{dyn} is proportional to the summation of all capacitance in the circuit and operating frequency, whereas P_{leak} is proportional to the supply voltage and leakage current. The most commonly

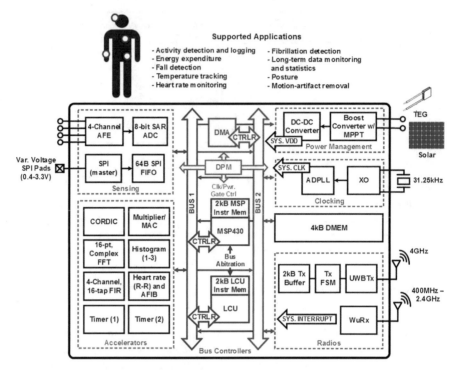

Fig. 3.11 Biomedical system for IoT [3]

used power reduction techniques in this respect are voltage scaling, power gating, and clock gating. Voltage scaling is an effective technique, as the dynamic power consumption is proportional to the square of the supply voltage. At low supply voltages, the leakage is also minimized. Hence, it is necessary to find an optimum supply voltage which gives the minimum energy point.

$$P_{Total} = P_{dyn} + P_{leak} \tag{3.11}$$

$$P_{dyn} \propto C_{Total} \times f \times V^2 \tag{3.12}$$

$$P_{leak} \propto I_{leakage} \times V \tag{3.13}$$

where, C_{Total}: total capacitance, f: operating frequency, V: supply voltage, $I_{leakage}$: leakage current.

Another important power saving mechanism is duty cycling. Wearable devices perform repetitive tasks that do not require the whole system to be always ON. In duty-cycled ultra-lower power systems, the average power consumption is given by Eq. (3.14) [51] and its operation is illustrated in Fig. 3.13. There are three components of the power consumption: the power dissipated by the always ON

Complete System Configuration

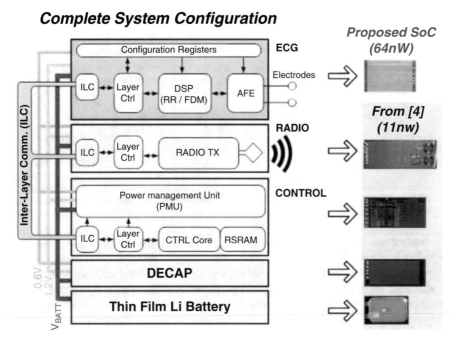

Fig. 3.12 Implantable ECG SoC [50]

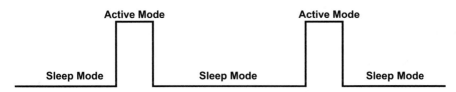

Fig. 3.13 Duty-cycled operation

block ($P_{always\text{-}on}$), the power in the sleep mode (P_{sleep}), and the active power (P_{active}). Highest power consumption occurs when all blocks are on during the active operation. In such duty-cycled systems there is a trade-off between ON-time, leakage power, and active power. Lowest energy operating point could be obtained depending on the design complexity and power dissipation of each part. Such duty-cycling is advantageous in IoT devices especially for transmitter since the transmitter is the most energy-hungry part and it does not have to be turned on always. Transmission could be done periodically.

$$P_{ave} = P_{always\text{-}on} + P_{sleep} + \frac{E_{active}}{T_{wkup}} \qquad (3.14)$$

Fig. 3.14 Duty-cycled
energy optimization [52]

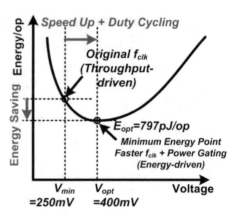

The advantages of duty-cycled operation [52] are illustrated in Fig. 3.14. It was applied to reduce the energy consumption of one block of the design. As the supply voltage is lowered both leakage and dynamic power reduce. In order to avoid leakage energy per cycle, it was proposed that the system operate at higher frequency than the minimum required number of cycles per task. The block that was optimized was performing arrhythmia detection and required to operate for 500 cycles every 10 s window. Though 500 Hz was sufficient, it was made to operate at 10 kHz in burst mode. Through such optimization the energy per operation was 797 pJ/op when operating at 400 mV.

3.6 Summary

In this chapter, several ECG processing techniques were reviewed. ECG processing algorithms for QRS detection, full ECG feature delineation, and ECG classification were briefly described. In addition, various biomedical SoCs for sensing various physiological signals have been reported. All of the recent biomedical SoCs employ one or more power reduction techniques such as voltage scaling, clock gating, and near threshold operation. In addition hardware accelerators aid in enhancing energy efficient operation.

Chapter 4
Combined CLT and DWT-Based ECG Feature Extractor

4.1 Introduction

Wearable ECG processing platform includes signal sensing, acquisition, local processing, recording, and transmission. All of these components have to be designed to fit on an integrated system fulfilling the stringent power and area requirement. Signal sensing utilizes electrodes to pick the signals from the human body while acquisition is achieved through amplifier and analog to digital converter.

Analyzing the acquired ECG signal requires efficient algorithms in order to extract vital features from ECG. Existing techniques in ECG processing included discrete wavelet transform (DWT) [29, 53] for full ECG feature extraction, time domain based on Pan and Tompkins technique (PAT) [24] for QRS detection, and curve length transform [54] for QRS detection. DWT could be utilized specifically for T and P wave detection as in [53]. DWT has good noise suppression capabilities; however, its hardware implementation requires huge memory in order to store multiscale resolutions. PAT implementation involves multiple operations such as differentiation, moving averaging, and squaring which have computational complexity. In this chapter, these issues will be addressed by optimizing computations and reducing memory requirement.

Optimization, both at an algorithmic level and an architectural level, enables ultra-low power consumption for on-chip implementation of such systems. Reported state-of-the-art ECG processors were capable of operating at $8.47\,\mu W$ for QRS detection [7] and at $5.97\,\mu W$ for full ECG feature extraction [8]. Low energy dissipation is obtained through this specialized configuration such as non-volatile memory configuration for QRS detection [7], or application of low-power techniques (e.g., voltage scaling, frequency scaling, or clock gating) [55].

This chapter presents the design and implementation of an ultra-low power ECG feature extraction engine as part of a self-powered integrated platform that utilizes a thermal energy harvester. Thermal energy harvesters have limited power in the order

© Springer International Publishing AG, part of Springer Nature 2019
T. Tekeste Habte et al., *Ultra Low Power ECG Processing System
for IoT Devices*, Analog Circuits and Signal Processing,
https://doi.org/10.1007/978-3-319-97016-5_4

of 20–40 μW. Hence the power dissipation has to be low enough (less than 10 μW) in order to be supported by thermal energy harvesters. Existing micro-controllers have an active power dissipation of greater than 100 μW [56], which makes them incapable of being powered by energy harvesters [57]. Moreover, the active power consumption is directly related to the energy dissipation and battery life, which enable the proposed system to fit in energy constrained wearable devices.

The performance of the overall system was validated using an ECG signal database (for real patients) acquired from the American Heart Association (AHA). Extracted features using the proposed processor could be utilized to determine the intervals in order to detect or predict cardiac diseases [12]. System description along with ECG feature extraction architecture and ultra-low power operation is described in Sect. 4.2. Section 4.3 shows the measurement results and analysis. Section 4.4 summarizes the chapter.

4.2 System Description

The system architecture consists of all the digital circuitry required for ECG feature extraction as depicted in Fig. 4.1. ECG, being a non-stationary signal, is corrupted by various types of noise: including power-line interference, baseline wonder, and motion artifacts [58]. All ECG processing systems comprise one or more noise suppression components. The proposed system has a bandpass filter to suppress noise. Following the ECG pre-processing, the main component in the system is ECG feature extraction. It is the task of the feature extraction engine to delineate each of the ECG wave components. The flow of data and all the operations of the digital circuitry are controlled through a custom finite state machine (FSM). Custom-designed FSM is much more power efficient when compared to a general-purpose processor-core. Moreover, the control FSM provides the necessary timing signals for the coordination of each stage and implements clock gating. The architecture for feature extraction (illustrated in Fig. 4.1) is constructed based on an algorithm in [59] and is described in the following subsection. In [59] the algorithm was implemented in Matlab using double precision; however, the proposed architecture uses fixed-point precision, which is fully optimized for hardware realization.

4.2.1 ECG Feature Extraction Architecture

The flowchart of the feature extraction process is shown in Fig. 4.2. It performs three functions: filtering, ECG transformation, and ECG delineation. Filtering enhances the SNR of the ECG wave and removes low-frequency artifacts, such as baseline wonder and motion artifacts as well as high frequency interference. Various ECG features have their own unique characteristics. For instance, the QRS complex forms a relatively higher amplitude, whereas the P and T waves are characterized by

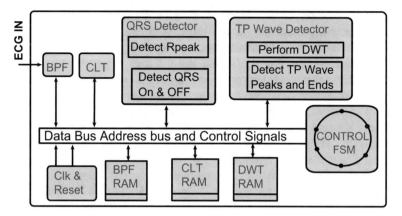

Fig. 4.1 ECG processing architecture

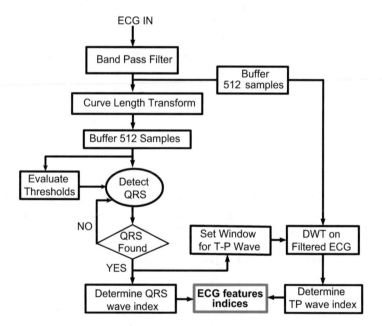

Fig. 4.2 Flow diagram of the ECG system

curved structures with relatively smaller deviation from the baseline. Detecting the QRS complex is a primary step in feature extraction process. Robust and accurate detection of QRS complex is followed by refined determination of T and P waves. In the proposed architecture, CLT is used for QRS detection, and DWT is utilized for T and P wave detection. CLT offers a computationally efficient QRS detection technique [54, 59].

The main features of ECG are contained in the frequency range 0.5–50 Hz, hence the sampling frequency of our proposed architectures is 250 Hz. ECG is oversampled above Nyquist rate to enhance detection accuracy. Digitized ECG samples with 8-bits are directly filtered and the filter is pipelined operating at a clock rate of 250 Hz and is clock gated at other time slots.

The system contains a buffer of 512 samples for both the filtered and CLT signal as shown in Fig. 4.2. The main purpose of the CLT buffer is to enable backward search and forward search for QRS onset/QRS offset points based on a window relative to the detected QRS peak. The buffer for the filtered signal is utilized for detecting T and P waves. Even though these buffers exist, the QRS detection is continuous fulfilling the real-time requirement of the ECG signal. The minimum heart rate that could be detected using our proposed system is 30 bpm (1 beat/2 s) and 2 s corresponds to 512 samples based on a sampling rate of 250 samples/s. Thresholds for QRS detection are evaluated with each new sample. Also a new sample is stored in the buffer even if the system is processing, since it only requires one clock cycle to filter, CLT transform and store a single sample. It requires 2.048 s in order to acquire 512 samples, and on average, 2000 clock cycles are required to process 512 samples so as to extract one full P-QRS-T wave. The required clock cycles vary according to the morphology of the incoming signal hence the operating frequency is set higher than the minimum necessary frequency to accommodate the variations. The proposed system was verified to operate at a minimum frequency of 7.5 kHz. The total energy required to perform 2000 cycles is 171.2 nJ.

4.2.1.1 QRS Detection

QRS detection was performed using CLT. CLT extracts slopes and lengths of successive points of a wave as given in Eq. (4.1). When CLT is applied to ECG, it enhances the QRS complex and relatively suppresses the TP waves. CLT signal obtained from chip test results for different ECG records is shown in Fig. 4.4. CLT has an advantage in that it handles various ECG morphologies and suppresses baseline drift.

A novel pipelined architecture is proposed for implementing the CLT as shown in Eq. (4.2). During each stage of the pipeline, only one square root function is required (the second term in Eq. (4.2)), whereas the other terms are obtained from preceding transforms. In each stage of the pipeline only one square root, one addition, and one subtraction are enabled. This pipelined architecture is illustrated in Fig. 4.3 where the first block evaluates the difference of successive samples followed by squaring and summation. The last two blocks perform square root and accumulation. The window size for the proposed CLT is 32. There are another two approaches for realizing the CLT **Approach 1**: Replicating the resources 32 times such as the squaring, square root, and addition, and **Approach 2**: Reusing the same resources for 32 clock cycles. Approach 1 requires 32× more resources than the proposed technique and consequently more power. Moreover, approach 2 requires 32× more clock cycles than the proposed technique. Table 4.1 summarizes synthesis results

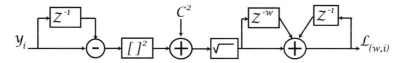

Fig. 4.3 Pipelined architecture for CLT

Table 4.1 Comparison of CLT realization techniques

CLT technique	Approach 1	Approach 2	Proposed technique
No. of combinational cells	33,498	1028	1028
No. of sequential cells	1985	598	598
Total no. of cells	35,483	1626	1626
Leakage power (μW)	1.47	0.10	0.10
Dynamic power (μW)	3.83	1.23	1.23
Power at 100 kHz (μW)	5.3	1.33	1.33
Energy per single transform (nJ)	0.053	0.4256	0.0133

for three approaches in terms of the required resources and energy dissipation. The proposed pipelined CLT requires only 7.4% of the total area need for Approach 1 and it requires 32× lower energy than Approach 2 in order to obtain one sample of the CLT. Moreover, the throughput is increased by 32× compared to Approach 2, eliminating the need for 32 clock cycles to obtain one sample of CLT output. The leakage power is reduced by 10× when comparing Approach 1 and the proposed technique due to the reduction in the required resources. Though Approach 1 uses parallel operation for implementing the CLT, its input is coming from a set of ECG samples stored in registers which requires pipelining the input samples. Hence, the throughput for Approach 1 and the proposed pipelined architecture is the same. For every new sample, there is one transformed output.

$$L(\omega, i) = \sum_{k=i-\omega}^{i} \sqrt{C^2 + (\Delta y_k)^2} \tag{4.1}$$

$$L(\omega, i) = L(\omega, i - 1) + \sqrt{C^2 + (\Delta y_i)^2} \\ - \sqrt{C^2 + (\Delta y_{i-1-\omega})^2} \tag{4.2}$$

$$Th_{up} = \frac{2}{3} \left[mean \ (CLT_{Pre_{beat}}) + Th_{pre} \right] \tag{4.3}$$

A sample CLT signal for two different ECG records is illustrated in Fig. 4.4. The transformed signals are similar though their respective original signal has variations (e.g., ECG with upward and inverted QRS). This makes it easier to develop a unified

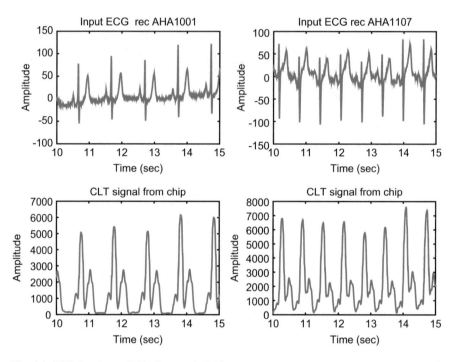

Fig. 4.4 CLT signal out of chip for sample ECG

method for QRS detection for wide range of ECG morphologies. The CLT signal, which is saved in memory, is used to evaluate the threshold required for the detection of QRS peaks. The threshold is adaptive to the incoming signal and is evaluated as in Eq. (4.3). After QRS_{peak} is detected, the system goes on to locate the Q_{on} and Q_{off} by defining a search window to the left and right of QRS_{peak}. Q_{on} and Q_{off} are also obtained by applying thresholds on the CLT signal.

4.2.1.2   DWT-Based T and P Wave Delineation

T and P are obtained by applying DWT on a window after and before the QRS complex on the filtered ECG, respectively. The search windows are adaptive and are updated based on the previously detected RR interval. DWT is implemented as a cascade of filter banks and provides a multiscale decomposition. A single scale 2^3 is used for the TP wave delineation process. Since the DWT is performed in a small window (less than 120 samples) and a single scale is used, the required memory is minimized relative to utilizing multiscale decomposition on the whole ECG signal.

The wavelet decomposition forms a pair of maxima based on the concave or convex nature of the TP waves around the baseline. This pair of maxima known as maximum modulus pair (MMP) is detected in order to locate the positions of the TP

Fig. 4.5 Clock gating

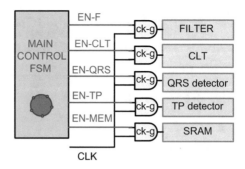

waves. First, a threshold is set as the mean of the absolute value of the DWT signal, and then a peak is detected followed by zero crossing and second peak. If the first peak is a maxima, then it is a concave signal around the isoelectric line, and if the first peak is a minima, then it is a convex signal. The zero crossing point sets the location of the T or P wave peaks. Their respective onset and offset are obtained by extrapolating the distance between the zero crossing of the DWT and the peaks to the left and right of the zero crossing.

4.2.2 Power Reduction Techniques

In the proposed design, clock gating is utilized to minimize the overall power consumption. Clock gating enables power saving by reducing the dynamic power and leakage. Since all the procedures are not performed simultaneously, some blocks could be clock gated. Figure 4.5 depicts the clock gating setup, and Fig. 4.6 depicts the timing diagram for the clock gating setup. The clock gates are controlled by the main control FSM. Signals EN-F, EN-CLT, EN-QRS, EN-TP, and EN-MEM correspond to enabling filters, CLT, QRS detector, TP delineation, and memory, respectively. Filtering is done whenever a new sample is acquired which is at the 250 Hz sampling rate, hence the filter is clock gated at all other time slots. CLT is also performed at the speed of the sampling rate since its pipelined architecture requires only one clock cycle to transform an incoming sample. QRS detection and TP detection are clocked at different time slots as in Fig. 4.5 since the TP wave detection depends on the QRS detection. Moreover, the clock to the SRAMs is enabled only during the write and read operations.

A total of 2237 bits out of 4037 bits that map into registers are utilized for analyzing the clock gating. These registers are contained in the filter, CLT, QRS, and TP blocks. Figure 4.7a shows the power consumption with and without clock gating when operating at 7.5 kHz. Power saving due to clock gating is illustrated in Fig. 4.7b. A maximum power saving of 34% is achieved when it operating at 0.7 V.

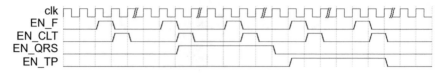

Fig. 4.6 Clock gating timing diagram

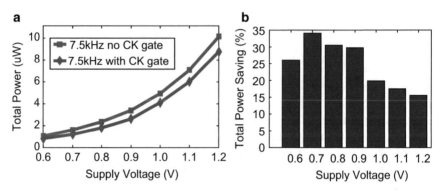

Fig. 4.7 (**a**) Total power with and without clock gating at 7.5 kHz. (**b**) Total power saving due to clock gating at 7.5 kHz

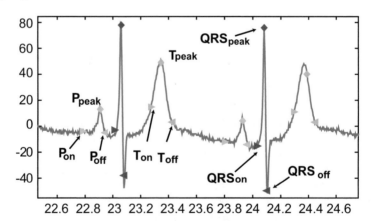

Fig. 4.8 ECG feature extraction example

4.3   Implementation and Measurement Results

The proposed architecture was implemented on chip using 65 nm low power technology. Figure 4.8 demonstrates the automatically extracted ECG features annotated on top of an input ECG signal. Figure 4.9 reveals the extracted intervals such as $P_{on} - R$, $R - T_{peak}$, and QT. Moreover, the figure shows the extracted

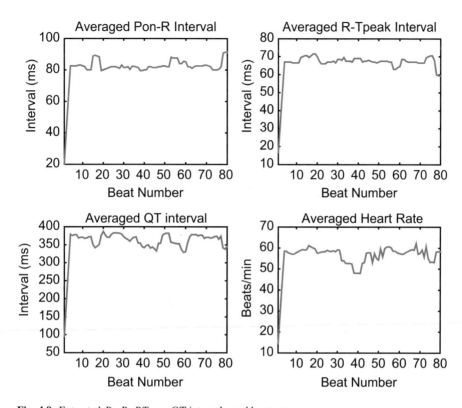

Fig. 4.9 Extracted $P_{on}R$, RT_{peak}, QT intervals, and heart rate

heart rate. These intervals, along with the heart rate, could be utilized for a classifier system, such as arrhythmia detection or VA prediction.

The proposed architecture is computationally efficient. Feature extraction is performed using CLT and DWT. The CLT is pipelined and requires only one square, one square root, and an addition for each sample. And the DWT is implemented as a cascade of filters. The filter coefficients are 2 and 1/2 which are implemented using shift and do not require multiplication or division. Besides, the DWT is performed on a small length of the signal on a window which is the probable location of TP waves. Only 256 bytes of memory is required for the DWT since only a single scale is used for the delineation process and the maximum window size does not exceed 256.

Prior to tapeout, the power consumption of the design was estimated using state-of-the-art chip design tools for synthesis and layout tools for digital circuits. For fabricated chips, the power was measured using DC-power supply that has integrated power analyses from Agilent. Figure 4.10 reveals the power consumption of the system when operating at a different frequency. The system could operate from a supply voltage of 0.6 V up to 1.2 V and is characterized at an operating frequency of 7.5–200 kHz. The power consumption is dominated by leakage, and if

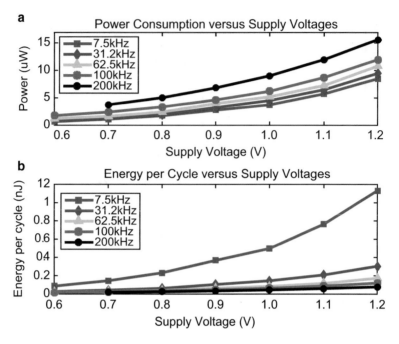

Fig. 4.10 (**a**) Measured power consumption. (**b**) Energy per cycle

Table 4.2 Leakage power

Supply voltage (V)	0.6	0.7	0.8	0.9	1.0	1.1	1.2
Leakage power (μW)	0.54	0.98	1.50	2.45	3.37	5.35	7.90

the leakage is reduced, then overall power consumption would reduce significantly. According to Fig. 4.11 and Table 4.2 the dynamic power is less than 40% of the overall power consumption. Even though the power is dominated by leakage, the energy per cycle is lower at higher frequencies which implies that a higher operating frequency leads to better energy efficiency. The proposed system is suitable to be powered from an energy harvester which could supply power in the order of microwatt range [57].

Performance evaluation results show that the QRS detector has a sensitivity of 98.5% and a predictivity of 98.2% when verified using the AHA database.

Figure 4.13 shows the die photo of the fabricated chip and summarizes the performance of the system. The presented system performance results in comparison to the state of the art is revealed in Table 4.3. Our proposed system achieves full ECG feature extraction, with a power consumption of only 642 nW, when operating at a supply voltage of 0.6 V and a frequency of 7.5 kHz. The power consumption is 2% of the power dissipation of the QRS detector reported in [6] though the proposed system does full feature extraction. In addition, it is less than 15% of the power dissipation of the systems [7–9]. Even though the power in these systems

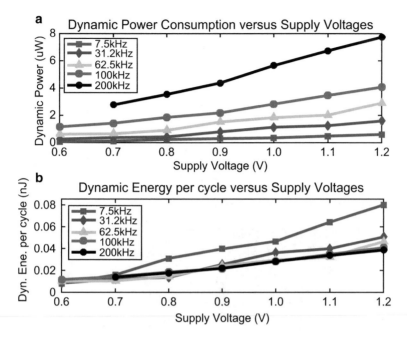

Fig. 4.11 (**a**) Measured dynamic power. (**b**) Dynamic energy per cycle

Table 4.3 Comparison with published work

	[6]	[7]	[8]	[9]	Proposed
Technology (nm)	65	130	180	350	65
Area mm^2	0.41	15.91	2.47	1.2	0.243
Oper. freq. (kHz)	7	32	0.12	1	7.5
Supply voltage	0.7	1.2	1.2	3	0.6
AFE	NA	IA	IA	SAR	NA
		ADC	ADC	ADC	
ECG features	QRS	QRS	QRS	P-QRS-T	P-QRS-T
Power (μW)	33	8.47	5.97	13.6	0.642
Energy per cycle (nJ)	4.714	0.265	49.75	13.6	0.0856

includes an instrumentation amplifier and ADC, their respective power consumption is dominated by their respective digital circuits. Moreover, the proposed system has the lowest average energy per cycle relative to [6–9]. In our proposed system, the energy required to process 512 samples of ECG is 171.2 nJ. However, the comparison is done based on the power consumption and the energy per cycle, since the papers that are used for comparison do not clearly indicate how many clock cycles are required to process a block of ECG.

Fig. 4.12 Energy optimization for duty-cycled operation

Technology	65nm
Supply Voltage	0.6 - 1.2V
Area	$0.234mm^2$
Operating Frequency	7.5 - 200 kHz
Power Consumption	0.642μW @ 7.5kHz, 0.6V

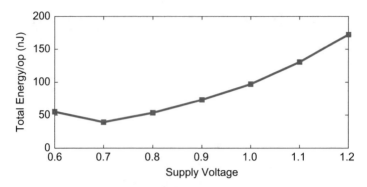

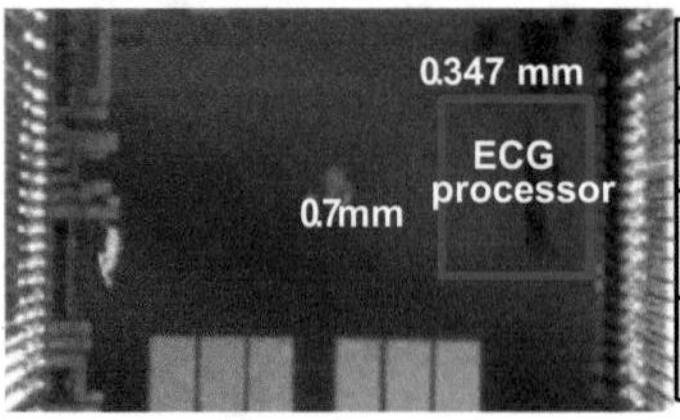

Fig. 4.13 Die photo and specification summary

In the fabricated chip, there were no power switches. Hence, it was not characterized for duty cycle operation. However, we can demonstrate how we can obtain the minimum energy based on the measured active and leakage power. This is illustrated in Fig. 4.12 where the chip is operating at 200 kHz for supply voltage greater than 0.7 and at 31.2 kHz at 0.6 V. Accordingly the optimum energy point will be at 0.7 V and at 200 kHz. Further analysis could be performed if we have a system that operates in the sub-threshold region. The supply voltage of the fabricated chip was limited due to the minimum operating voltage of SRAMs (Fig. 4.13).

4.4 Summary

An ultra-low ECG feature extraction engine was presented. The overall architecture was optimized for ultra-low power operations through the application of efficient computations and clock gating. A CLT-based QRS detection and DWT-based TP wave delineation were introduced. The system was fabricated using GF-65 nm low power technology and consumed 642 nW only when operating at a frequency of 7.5 kHz from a supply voltage of 0.6 V. The presented engine is suitable for integration in SoCs for wearable biomedical devices powered by energy harvesting.

Chapter 5
ACLT-Based QRS Detection and ECG Compression Architecture

5.1 Introduction

Ultra-low power medical devices are imperative in the era of the IoT. Healthcare sensors capture vital physiological data for monitoring and diagnosing patients. Holter monitors is a case in point where it records and monitors continuous ECG data for 24 h. They are constrained by power consumption since they need to operate for an extended period continuously. On the other hand, IoT healthcare platform enables minimum local processing and transfers data to cloud-connected servers that help resolve drawbacks of Holter monitors and similar devices. Cloud platforms provide easy access for doctors to continuously follow up on their patients. Various platforms of IoT architectures for healthcare were proposed as in [60, 61]. IoT healthcare connects patients, doctors, and devices according to the philosophy shown in Fig. 5.1.

IoT infrastructure extends from sensors, communicating devices up to central servers which incorporate efficient devices [62]. IoT platform challenges result from system engineering that involves signal acquisition, local processing, transmission, central processing, and generating feedback [63]. Each of these stages has challenges, especially with increasing numbers of connected devices.

ECG is one of the most vital signals in IoT healthcare devices. ECG, which represents the electrical activity of the heart, is used as a prime tool to monitor and diagnose cardiac diseases due to the non-invasive nature of ECG sensors and the accuracy of mapping between ECG signals and heart physical activity. ECG is utilized in cardiac arrhythmia prediction and detection by extracting ECG intervals, amplitudes, and wave morphologies of the different components such as the P, QRS, and T waves [23]. The basis for extracting such parameters depends on the accurate real-time delineation of the ECG wave components. The development of real-time and accurate delineation methods is crucial for abnormal ECG signals that occur with different types of cardiac diseases.

© Springer International Publishing AG, part of Springer Nature 2019   39
T. Tekeste Habte et al., *Ultra Low Power ECG Processing System
for IoT Devices*, Analog Circuits and Signal Processing,
https://doi.org/10.1007/978-3-319-97016-5_5

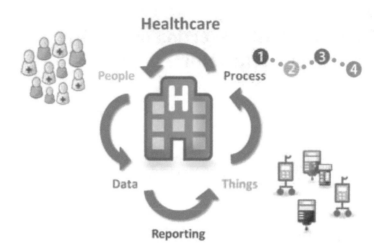

Fig. 5.1 IoT healthcare platform

The QRS complex, which is a principal component of the cardiac cycle, is used as a reference and represents the depolarization of ventricles in the heart. Its amplitude rises to 1 or 2 mV above or below the isoelectric line for normal beats and can go several times larger for abnormal beats. The time required for the ventricles to depolarize defines the QRS width or interval where it typically lasts between 80 and 120 ms [22]. QRS detection is a key for automatic delineation techniques. Various signal processing of QRS detection techniques have been proposed in the literature. Time domain thresholding along with filtering (first derivative, second derivative, both derivatives, matched filter, etc.) are some of the techniques that are suitable for real-time implementation [24–26]. In [24] Pan and Tompkins algorithm (PAT), which is one of the most widely researched and implemented techniques, was proposed, since it is robust in detecting QRS [12, 64]. Other methods that provide enhanced accuracy are based on the spectral analysis of the ECG signal. In [27–30], wavelet transform is presented as a tool to analyze ECG signals. As a part of the spectral analysis techniques, discrete Fourier transform has been reported in the literature to detect the QRS complex [31]. Empirical mode decomposition and Hilbert transform have been introduced to improve the analysis of the QRS detection of nonlinear and non-stationary ECG signals [32, 33].

Processed ECG data or extracted features in the IoT platform are transmitted wirelessly. Wireless data transmission is the most energy-hungry part in IoT devices. One of the effective ways in reducing energy consumed in wireless transmitters is to reduce the data transmitted through data compressors. In healthcare applications, lossless compression during transmission is a primary choice for reliability issues. Lossless ECG compressor architectures were reported in [65, 66]. Some recent data-compression schemes focused on lossy compression since it provides a high compression ratio [67]; however, it is less reliability when compared to lossless techniques. Lossy techniques have a high compression ratio in the range $2\times$ up to $15\times$. However, lossless compressors provide a compression ratio range of $1\times$ up to $3\times$.

Another option of reducing transmitted or processed data is decreasing the number of samples. In [68] a non-uniform time sampling technique is proposed with an adaptive sampling rate to reduce the energy consumption of the sampling process. Such a scheme is applicable to slowly varying signals. In [69] compressed sensing is presented as a potential technique for reducing the sample count, which is advantageous in reducing the overall power dissipation.

General-purpose micro-controllers could be the central processing unit of an IoT device. However, existing micro-controllers have an active power dissipation of greater than $100\,\mu W$ and a leakage power of greater than $1\,\mu W$ [70, 71], which is much higher power dissipation than that of custom ASIC solutions. Henceforth, the main reason to have a custom HW solution is to enable ultra-low power operation. The objective of this chapter is to present an ECG processing and compression architecture that will help IoT medical devices to achieve ultra-low power operation and to minimize the data needed to be transmitted to minimize power consumption. Operating at an ultra-lower power would enable the device to be powered by an energy harvester that generates power in the order of μW [72]. In this chapter, a multiplier-less ECG QRS detection architecture, which is based on a single transformation, is presented. Moreover, a compression technique based on first-derivative is proposed. The proposed QRS detection architecture consumed a $6.5\,nW$ when implemented in 65 nm low-power process.

The remaining part of the chapter is organized as follows: Section 5.2 provides a summary of existing QRS detection techniques, Sect. 5.3 contains the full description of the proposed QRS detection architectures, Sect. 5.5 presents performance evaluation and results, Sect. 5.6 discusses the compressor comparison with literature, and Sect. 5.7 concludes the chapter.

5.2 Summary of QRS Detection and Compressor Architectures

5.2.1 Summary of QRS Detection Architectures

QRS detection is challenging due to the following reasons. ECG (being low amplitude in nature) is contaminated by noise and artifacts, such as electrode noise, motion artifacts, muscle noise, power-line interference, ADC quantization noise, and noise in acquisition devices. Moreover, QRS waves have wide morphological variations among different people with different health conditions. Several QRS detection architectures have been reported in literature each having its own merits and demerits. Here are some of the commonly existing architectures.

A. Discrete Wavelet Transform QRS detection based on quadratic spline wavelet transform is reported in [73]. Even though the system achieves high sensitivity and predictivity (99.31% and 99.7%) for QRS detection when validated using MIT-BIH database, its implementation is so complex that requires scale-3 wavelet transforms and maximum modulus recognition. Its operating power consumption is $0.85\,\mu$W.

B. Differentiation and Adaptive Thresholding In [64] a QRS detection archi-tecture (QRS detection using differentiation, moving average, and squaring) is reported. Dynamically adaptive thresholds are applied to a squared ECG signal in order to detect QRS peaks. The system is optimized for an ultra-low power application that reduces computational complexity, however, still uses hardware-intensive operations such as multiplication and division.

5.2.2 Summary of ECG Compression Architectures

Several ECG compression architectures have been proposed and some of them are summarized below.

Fan Architecture Fan architecture for lossy ECG compression is reported in [74]. Fan is initially proposed in [75]. It operates by drawing the longest possible straight line between the starting sample and the ending sample, in such a way that during the reconstruction of samples, the error is less than the maximum specified error value, ϵ.

Lossless-Compressor Based on Linear Slope Predictor A low-power ECG compressing architecture, based on linear slope predictor, is reported in [65]. Moreover, it includes a fixed-length packaging-scheme for serial transmission. The architecture was implemented in $0.35\,\mu$m technology and achieves a compression ratio of $2.25\times$, at a power consumption of $535\,$nW, from a supply of $2.4\,$V for ECG sampled at $512\,$Hz.

Lossless-Entropy Encoder with Adaptive Predictor The system in [66] presents a unique lossless ECG encoder based on an adaptive rending predictor and two-stage entropy encoder. When the design was synthesized in $0.18\,\mu$m technology, it consumed $36.4\,\mu$W at operating frequency of $100\,$MHz. It achieved a compression ratio of $2.43\times$.

5.3 Proposed QRS Detection Architecture

The overall block diagram of the proposed ACLT architecture, along with the compressor, is illustrated in Fig. 5.2. In this chapter, the main contribution is in the QRS detection architecture and compressor. Even though the ultimate goal of the compressed data is to be transmitted wirelessly, issues related to the transmitter

Fig. 5.2 Block diagram of proposed algorithm

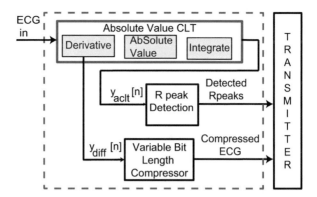

such as transmission error are beyond the scope of this chapter. However, in IoT devices, it is necessary to quantify the packet error rate with regard to the signal-to-noise ratio of the wireless transmitter [76].

QRS detectors should be robust enough to deal with the noise and artifacts mentioned in the previous section. It is challenging to come up with a generalized system that deals with all the artifacts at the same time. Filtering has been widely used especially for removing low-frequency noise, baseline drift, and high-frequency interference. Transformation is applied to enhance a portion of the ECG waves. Our proposed system provides optimized QRS detection architectures that could deal with all the artifacts with minimum hardware resources without compromising accuracy.

5.3.1 Algorithm Formulation

Conventional ECG processing flow consists of pre-processing, transformation, and thresholding. Each of these stages requires huge computation in filtering and enhancing ECG. In this proposed technique, the pre-processing and transformation are lumped into one component, forming a modified version of curve length transform (CLT). CLT was reported in [54, 59] and it offers a computationally efficient QRS-detection technique.

CLT, for a discrete signal y_i over a time window ω, is given in Eq. (5.1). Equation (5.1) is referred to as conventional-CLT (C-CLT) in this chapter. The CLT can be re-written and evaluated as in Eq. (5.2). The symbol Δi^2 corresponds to the square of the sampling period (which is a constant value) and replacing it with a nonlinear scaling factor C^2 adds flexibility to manipulate the length-response ratio. C^2 is determined experimentally, taking into account the window size and the maximum height of the QRS complex. By choosing a proper value for it, a particular portion of the signal is improved and boosted in comparison to the rest of the signal.

$$L(\omega, i) = \sum_{i-\omega}^{i} \sqrt{1 + \left(\frac{\Delta y_i}{\Delta i}\right)^2} \Delta t \tag{5.1}$$

$$L(\omega, i) = \sum_{i-\omega}^{i} \sqrt{C^2 + \Delta y_i^2} \tag{5.2}$$

As shown in Eq. (5.2), the CLT integrates successive lengths over a fixed window. Hardware realization of Eq. (5.2) would require addition, multiplication, and calculation of the square root. In order to minimize the resources, Eq. (5.2) could be reformulated as in Eq. (5.3) where the square root is removed. In this chapter, Eq. (5.3) is referred to as squaring-CLT (S-CLT).

$$L(\omega, i) = \sum_{i-\omega}^{i} C^2 + \Delta y_i^2 \tag{5.3}$$

Furthermore, Eq. (5.3) is modified to form Eq. (5.4) where absolute value function replaces the squaring. Hence in this approach both the square and square root functions in Eq. (5.2) are replaced by the absolute value function. This becomes an absolute-value-CLT (ACLT). A multiplying factor 4 is added in Eq. (6.3) to relatively enhance higher ECG slopes and suppress noise which is centered at the baseline. Multiplication by a factor of 4 is implemented as shifting in hardware realization. Using this approach, we are minimizing the resources that would be required to implement the CLT. Its performance and required hardware resources, with respect to other approaches, will be discussed in Sect. 5.3.

$$L(\omega, i) = \sum_{i-\omega}^{i} \left| C^2 + |4 \times \Delta y_i| \right| \tag{5.4}$$

All of the above three approaches (Eqs. (5.2), (5.1), (5.4)) could be applied for QRS detection as the CLT also has an inherent behavior for suppressing the baseline wander of ECG. Based on the above analysis, the CLT could be evaluated using these three approaches, namely: (1) conventional CLT, (2) squaring-CLT (S-CLT), and (3) absolute-value CLT (ACLT). Figure 5.3 shows the transforms for ECG data from MIT-BIH record 112, where the signals have baseline wandering. Though all of the three approaches are feasible, in this chapter, only C-CLT and ACLT are implemented and compared since S-CLT has a large amplitude range about the other two approaches to such a degree that its hardware realization would require more bit width. Also, S-CLT has poor performance in suppressing baseline wander as could be observed in Fig. 5.3, and consequently, its detection accuracy was low. The detailed architecture of the ACLT is presented in the next subsection.

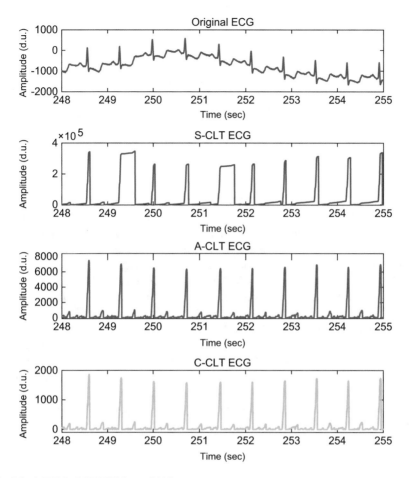

Fig. 5.3 ACLT for MIT-BIH Record 112

5.3.2 Proposed ACLT Architecture

Figure 5.4 shows the proposed ACLT architecture for detecting QRS complex. It is an architecture for the algorithm formulated in Eq. (5.4). It performs transformation followed by QRS peak detection using adaptive threshold. The transformation is done using derivative, absolute value, and integration (all lumped into one realization of the ACLT). The transformation distinctively enhances QRS complex even for noisy ECG signals corrupted with baseline wander. Its uniquely inherent behavior removes the need for additional complicated circuits for high-pass or low-pass filters. All of the computations for the transformation are performed using addition and shifting. Moreover, comparison is required for detecting QRS peaks using thresholds. There is no need for multiplication, division, or square root function. Hence its hardware implementation requires only adders, shifters, and

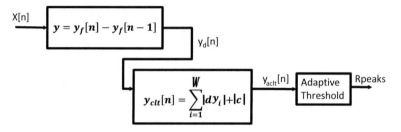

Fig. 5.4 Proposed absolute-value-CLT

comparators. These components are less hardware intensive relative to multipliers, dividers, and square root functions. For instance, if we compare an N-bit multiplier with an N-bit adder, an N-bit multiplier would require N times N-bit adders. Alternatively, a multiplier would need N-times clock cycles. Division and square root are much more complicated than addition or shifting.

The integration over a window in the proposed architecture is pipelined. Pipelining enables it to transform directly whenever there is a new ECG sample. Accordingly, the required clock frequency for the architecture is equal to the sampling frequency of the incoming ECG signal. The sampling frequency of the system is 250 Hz. This is the lowest operating frequency possible for such a configuration. Such a low operating frequency reduces the dynamic power dissipation. Depending on the proposed architecture duty cycling would not give advantage since the design is operating at the sampling rate of the incoming ECG signal. Buffering the ECG signal and then processing at higher frequency would require buffers (SRAM) which add more leakage to the design.

5.3.3   QRS Peak Detection

QRS detection is performed using adaptive threshold. Applying threshold has been commonly used in detecting QRS peaks. However, it is necessary to construct an optimized technique to evaluate the thresholds. A threshold technique where the threshold is set to a mean of all previously detected R_{peaks} is reported in [64]. This threshold is updated according to Eq. (5.5) with every new sample, where the threshold factor P_{Th} is given by Eq. (5.6). The previous threshold is multiplied by a factor with every new sample. Even though using this adaptive threshold produced sensitivity and predictivity above 99%, it requires multiplication with every sample.

$$Th_n = Th_{n-1} * e^{\frac{-P_{Th}}{fs}} \tag{5.5}$$

$$P_{Th} = \frac{0.7 * Fs}{128} + 4.7 \tag{5.6}$$

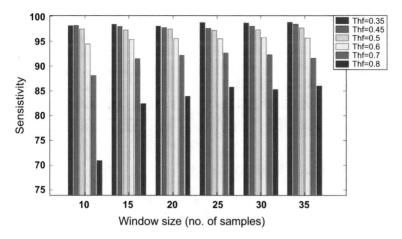

Fig. 5.5 Window and threshold factor selection

In our proposed architecture, the threshold is evaluated based on the equation given in Eq. (5.7). The threshold is updated whenever a new beat is detected and is proportional to the mean of the previously detected QRS peaks. Only eight previously detected QRS peaks are utilized in this stage. In hardware realization, division by 8 is implemented using shifting. The most challenging part in this step is finding the appropriate threshold factor to handle wide morphologically variant ECG waves from different standard databases. Many experiments were done using the standard database from Physionet in order to obtain optimum threshold factor. Figure 5.5 shows the effect of threshold factor on the sensitivity of QRS detection. The experiment was done on MIT-BIH. It is observed that, for a fixed window size of the ACLT, the sensitivity improves with as the threshold factor decreases. Further reduction of the threshold factor would lead to misdetection in which noise or T wave of ECG would be detected as QRS peaks. Figure 5.6 demonstrates the ACLT, along with the threshold, for record 112 from MIT-BIH ECG database.

$$Th_i = Th_{factor} * mean \sum_{k=i-8}^{i} Rpeaks_k \qquad (5.7)$$

Once a threshold is defined, the next step is to find a peak in the ACLT signal within a window in which the signal is greater than the threshold. Figure 5.7 shows the FSM that is developed to detect the QRS peaks. State 1 checks if the incoming ACLT signal is greater than a pre-calculated threshold. Initially, the threshold is set to half of the first maximum value of the first 2 s of ECG data. Then the threshold is updated by accumulating newly detected beats, as discussed above, according to Eq. (5.7). When the ACLT signal crosses the threshold value, the system goes on to state 2. In state 2, the system finds the maximum values in a window where the signal is greater than the threshold value. This max value is set as the location of

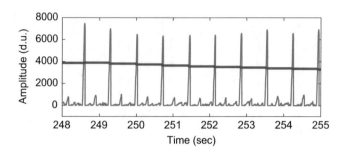

Fig. 5.6 Threshold value for record MIT-BIH 112

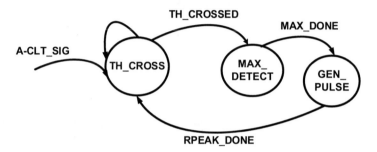

Fig. 5.7 QRS detection FSM

the QRS peak. State 3 generates a pulse indicating the detection of a new beat. This pulse is a fixed offset from the max value obtained in state 2 since the system has to check the whole window for locating the max value. After this, the system goes back to detecting the threshold crossing.

5.3.4   Optimization Parameters

According to the proposed architecture, there are two parameters that need optimum selection. These are the window size (w in Eq. (5.4)) and the threshold factor (Th_{factor}) in Eq. (5.7). In order to set these parameters, the sensitivity and predictivity of the resulting QRS detection are evaluated. Figure 5.5 shows the effect of window size and the threshold factor on the sensitivity of QRS detection. Note that (for a fixed window) the threshold factor has a major impact on Se. For a fixed threshold less than 0.6, the window size does not have much impact on Se. Based on this analysis, a window size of 15 and a threshold factor 0.375 are chosen. Threshold factor 0.375 is $1/2 + 1/8$, so in hardware realization, multiplication by 0.375 is implemented by shifting.

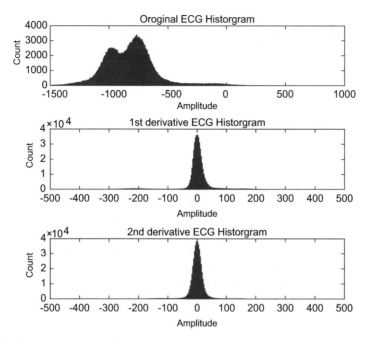

Fig. 5.8 Histogram for data distribution MIT-BIH rec1112

5.4 Proposed ECG Compression Architecture

A novel compression technique based on derivative is proposed. The system takes the first derivative and does a variable bit length compression on the *first derivative* signal. The reason the *first derivative* was chosen is that values from *first derivative* as well as from *second derivative* are concentrated around zero, as shown in Fig. 5.8. However, the amplitude of the original ECG is large amplitudes due to the fact that the QRS complex and its values are concentrated around the baseline. As a consequence more bits would be required to represent the original ECG than were necessary for the *first derivative*.

Our objective is to design an ultra-low power compressor that requires minimum hardware resource. The *first derivative* requires only adders. Moreover, the variable bit length encoder requires comparators or a priority encoder which could be easily implemented using combinational logic. Figure 5.9 shows the compressor architecture. However, the *first derivative* would be shared with the ACLT. There will be no additional hardware required to compute the *first derivative*. Figure 5.10 illustrates the flow chart for variable length encoder. A lesser number of bits are used for low-amplitude signals, and greater number of bits are used for large amplitude signals. Such encoding reduces the total number of bits required to represent the whole ECG signal, since the *first derivative* values are concentrated around zero.

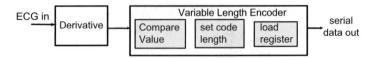

Fig. 5.9 Proposed compressor architecture

5.5 Performance and Results

To evaluate the performance of the algorithms, manually annotated ECG signals from Physionet MIT-BIH Arrhythmia Database and QT database are used [77]. MIT-BIH database contains 30-min-long, 48-two lead-ECG records sampled at 360 Hz, while the QTDB contains 15-min-long, 105-records, out of which 75 contain annotations for the QRS peaks. QTDB contains a wide variation of ECG data collected from other databases [78]. The proposed system was evaluated using the 48 records from MIT-BIH and 75 records from QTDB. MIT-BIH database contains randomly selected subjects as well as subjects with known arrhythmia that have clinical significance [79]. Moreover, the subjects are both men and women aged between 22 and 89 years. It has been widely used as a standard database for evaluating ECG QRS/arrhythmia detectors.

5.5.1 QRS Detection Performance

The proposed QRS detection architecture could detect various ECG morphologies including those with baseline wander, motion artifact, and noise corruption. Figure 5.11 shows ECG record 112 from MIT-BIH annotated with reference annotations (green) and detected annotations (red).

The performance of QRS complex detectors is evaluated before it is used in medical devices. The performance metric used in standard procedures is the sensitivity (Se) and positive predictivity (P^+). Detected QRS peaks are compared with reference annotation from experts. The sensitivity and positive predictivity are defined by Eqs. (5.8) and 5.9, respectively, where TP stands for the number of truly detected beats, FN denotes the number of false negative detection in which a beat exists but is not detected, and FP refers to the number of false-positive detection in which a beat does not exist but is detected.

$$Se = \frac{TP}{TP + FN} \times 100 \tag{5.8}$$

$$P^+ = \frac{TP}{TP + FP} \times 100 \tag{5.9}$$

Fig. 5.10 Variable length compressor flow chart

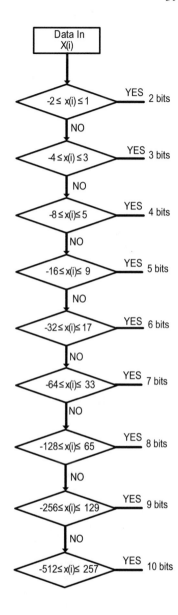

The detection performance obtained by the self-adaptive QRS detectors implemented in this work and other published detectors including [25, 27, 28, 73] and [64] are displayed in Table 5.1. The overall sensitivity of the implemented QRS detectors (based on C-CLT and ACLT) is found to be at levels of 99.0% and 99.37%, respectively. Following the same order, the positive predictivity is 99.3% and 99.38% when evaluated against the annotated beats in MIT-BIH. Table 5.1 shows that proposed ACLT performs well in the order of greater than 99.3% though

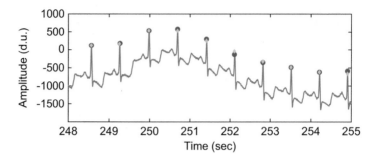

Fig. 5.11 QRS detection for MIT-BIH record 112

Table 5.1 Sensitivity and positive predictivity of QRS complex detectors (MIT-BIH)

Technique	Se	P^+
[25]	99.69%	99.77%
[27]	99.8%	99.86%
[28]	99.63%	99.89%
[73]	99.31%	99.7%
[64]	99.54%	99.74%
C-CLT	99.0 %	99.33 %
ACLT	99.37%	99.38%

its implementation is much less complex than that of the other referenced systems. Systems reported in [27], [28], and [73] are based on wavelet transform that requires multiscale decomposition which is implemented using FIR filters [64].

5.5.2   Computational Complexity of QRS Detector

Computational complexity gives a measure to evaluate the system for its suitability in ultra-low power IoT systems. For comparison, we have implemented three versions of the CLT. Moreover, we have made a comparison with the system implemented in [64] along with PAT as implemented in [64]. PAT is a widely reported QRS detection technique.

The computational complexity of the proposed algorithm is measured using the number of multipliers, adders, and comparators needed for the design. Table 5.2 reveals the resources required for the proposed architecture. The main superiority of the proposed ACLT architecture is that it does not require any multipliers. Though the number of adders and additions per second required in the proposed system is greater than in [64], the total operations per second is less than 50%. The proposed system requires 35% of the comparators required in [64] and 53% of the PAT implemented in [64].

Table 5.2 Resource consumption

Technique	[64]	PAT as in [64]	Conventional CLT	Proposed ACLT
Memory cells	28	123	18	18
Multipliers	6	6	1	0
Adders	5	41	13	13
Comparators	–	–	3	3
Square root	–	–	1	–
Square root./s	–	–	250	0
Mult./s	1107	1201	250	0
Adds./s	1205	1107	1261	1261
Comp./s	2163	1416	750	750
Total Ops./s	4475	5434	2512	2012

Table 5.3 Hardware resources and power

Technique	Conventional CLT	Proposed ACLT
Combinatorial cells	1082	657
Sequential cells	341	445
Buffers/inverters	101	146
Total cells	1423	1102
Area μm^2	13,940	10,074
Operating frequency	250 Hz	250 Hz
Leakage power	7.3 nW	5.16 nW
Dynamic power	1.6 nW	1.34 nW
Power	8.9 nW	6.5 nW

Relative to the C-CLT implemented by the authors, the proposed system does not use squaring or square root functions. Both squaring and square root functions (especially square root) are hardware-intensive operations. This implies that there is a 100% saving in multiplications and square root by implementing ACLT. Moreover, there is a saving of 27% on the power consumption as revealed in Table 5.3. Even though these operations are removed in order to attain ACLT, the performance is comparable. Even if we compare their sensitivities, ACLT achieves better results. If we look at Table 5.1, the proposed ACLT has a sensitivity and predictivity greater than 99%.

5.5.3 Compression Architecture Performance

Our proposed compressor is based on variable bit length for the first derivative of an ECG signal. Figure 5.12 illustrates a sample ECG and its first derivative. Relative to the original ECG, the signal amplitude range is reduced by a factor of 2. In addition,

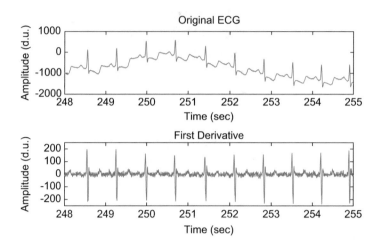

Fig. 5.12 Compression: first derivative for MIT-BIH record 112

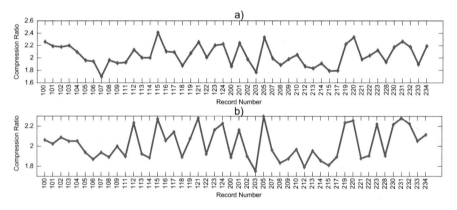

Fig. 5.13 Compression ratio for MIT-BIH (**a**) using first derivative and (**b**) using second derivative

the values in the first derivative are concentrated around zero; though, the original ECG has baseline drift.

The bit compression ratio is evaluated as in Eq. (5.10) where a total number of bits of uncompressed samples corresponds to the product of the number of samples with a fixed number of bits per sample (Eq. (5.11)). The MIT-BIH is sampled using 11 bits/sample. The number of bits of the compressed data corresponds to the summation of all bits from each sample (Eq. (5.12)). An average compression ratio of $2.05\times$ and $2.10\times$ is attained using the first and second derivative of ECG from MIT-BIH (as illustrated in Fig. 5.13). Figure 5.13 presents the compression ratio for all the records from MIT-BIH database. The compression ratio for all records is illustrated because the records have different morphologies and represent various cardiac conditions. Hence, it is a verification that the compression algorithm could handle various morphologies at a small range of compression ratio (within $1.7\times$ and $2.4\times$).

$$BCR = \frac{\text{T. No. of bits uncompressed samples}}{\text{T. No. of bits compressed samples}} \qquad (5.10)$$

$$\text{Total number of bits uncompressed samples}$$
$$= \text{No. samples} \times (\text{bits/sample}) \qquad (5.11)$$

$$\text{Total number of bits compressed samples}$$
$$= \sum \text{All bits of each sample} \qquad (5.12)$$

5.5.4   Hardware Implementations and Synthesis Results

The proposed architecture is coded using Verilog and simulated for functional verification. Its realization schematic is shown in Fig. 5.14. The design was synthesized using state-of-the-art tools from Synopsys, and layout was also generated. The standard cell library was fully characterized in silicon and is in an industry-standard tape-out-ready form. The standard cells were three flavors: LVT, RVT, and HVT. Though LVT cells have high leakage, they are more suitable for high speed applications and HVT cells for low-leakage applications where speed is not a major concern. RVT lies between LVT and HVT in terms of leakage and speed. The implementation was done using HVT cells, as HVT cells have more than $10\times$ lower leakage than RVT cells in 65 nm, in addition to the design being operated at low frequency. Post-layout power analysis shows that the ACLT system consumed a total power of 6.5 nW when operated from a supply of 1 V at an operating frequency of 250 Hz. The leakage power is 5.16 nW, accounting for 79% of the total power. The leakage power could be optimized by powering from a lower supply voltage, and the system could go up to 0.4 V for 65 nm technology [52]. We can estimate leakage saving at lower voltages, as the leakage is linearly related to the supply voltage. For instance, if the leakage at 1 V is 5.16 nW, then the leakage will be 2.064 nW at 0.4 V (which is a reduction of 60% in the leakage power).

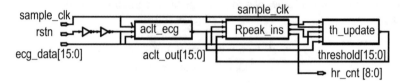

Fig. 5.14 Schematic ACLT core

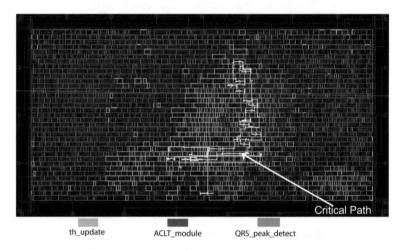

Fig. 5.15 Layout of the ACLT core

The layout of the proposed ACLT architecture is revealed in Fig. 5.15 which was generated using IC compiler from Synopsys. Design hierarchy and the worst case timing path are annotated in the figure. Timing verification was also performed, and the design has positive slack meeting all setup and hold time requirements. Timing closure was achieved using design constraint based on the standard cell characteristics.

5.6 Compressor Comparison with Literature

Table 5.4 shows the comparison of the proposed compressor with literature. The proposed lossless compression architecture consumed only 3.9 nW when operating at a frequency of 3 kHz, at supply voltage 1 V. The leakage is 0.51 nW, accounting for 13.1%. Operating frequency is set to 3.0 kHz so as to transmit the maximum number of bits serially from the variable length encoder within the sampling time of the input ECG signal. Even though the proposed architecture has a compression of 2.05, slightly lower than that reported in [65, 66], its implementation only requires 0.179 K gates and only consumes 3.9 nW. The system in [66] (being a standalone compressor) consists of a predictor followed by the entropy encoder. However, the compressor in [65] is part of a complete ECG processing system that includes an analog front end. Since the compressor subsystem performance (power and area) are reported separately, these metrics are used for comparison. Therefore, the comparison that is reported in Table 5.4 is apple-to-apple.

Table 5.4 Compressor comparison with published work

	[66]	[65]	Proposed
Technology (nm)	180	350	65
Oper. freq.	100 MHz	32 kHz	3 kHz
Supply voltage	1.8	2.4	1
Compression ratio	2.43	2.25	2.05
ECG channels	1	1	1
Total gate count	3.57 K	2.26 K	0.179 K
Power (μW)	36.4	0.535	0.0039

5.7 Summary

This chapter presented a real-time QRS detector and ECG compression architecture for energy constrained IoT healthcare wearable devices. An ACLT that effectively enhances QRS complex detection with minimized hardware resources was proposed. The proposed implementation required adders, shifters, and comparators and avoided the need for any multipliers. QRS detection was accomplished using adaptive thresholds in the ACLT transformed ECG signal. The proposed QRS detector achieved a sensitivity of 99.37% and a predictivity of 99.38% when validated using databases acquired from MIT Physionet. Furthermore, a lossless compression technique was incorporated into the proposed architecture using the ECG signal first derivative and variable bit length, an average compression ratio of 2.05 was achieved when evaluated using MIT-BIH database. The proposed QRS detection architecture was implemented using 65 nm low-power process; it consumed an ultra-low power of 6.5 nW when operated at a supply of 1 V. Also, the proposed compressor consumed only 3.9 nW when operated at a supply of 1 V.

Chapter 6
Ultra-Low Power CAN Detection and VA Prediction

6.1 Significance of CAN

Type 2 diabetes mellitus is one of the major prevalent diseases worldwide, and it is also one of the main causes of cardiovascular abnormalities leading to increased morbidity and mortality. Cardiac autonomic neuropathy (CAN) is a complication of diabetes and is characterized by an abnormality in the associated cardiac rhythm. Furthermore, CAN is often associated with other diabetes related complications such as organ dysfunction including altered sudomotor dysfunction, pupillary reflexes, gastroparesis, exercise intolerance, sexual dysfunction, and impaired neurovascular function [49–80].

Cardiac autonomic reflex testing (CARTs) were used to diagnose CAN and progression of CAN as proposed in [81]. These are non-invasive and consist of five tests that include heart rate responses stimulated by controlled breathing, the Valsalva maneuver, standing from seated or supine position, and the changes in blood pressure induced by standing and forced handgrip. CARTs are time consuming and in many instances counterindicated for use in general screening and are therefore mostly performed in hospitals or healthcare centers. It is also not possible to use CARTs with mobile device.

CARTs are the standard procedure for determining the normal procedure for determining CAN progression. However, CARTs also have limitations that require the active participation of the patient. This is not always possible due to the loss of mobility in older or fragile patients, those with cardiorespiratory pathology or obesity [82]. Methods that can diagnose CAN using only surface ECG require minimum patient interaction and are increasing in popularity. In [82], QT variability alone independent of HR variation was proposed for detecting CAN.

Heart rate variability (HRV) is commonly analyzed relative to frequency domain or time domain measures as well as nonlinear measures including the fractal dimension using information derived from ECG datasets. Heart rate variability

© Springer International Publishing AG, part of Springer Nature 2019
T. Tekeste Habte et al., *Ultra Low Power ECG Processing System for IoT Devices*, Analog Circuits and Signal Processing,
https://doi.org/10.1007/978-3-319-97016-5_6

analysis is used to detect arrhythmia which might lead to adverse cardiac events. Interbeat interval differences which are related to heart rate can be assessed using Shannon entropy and the more generalized Renyi entropy. Renyi entropy can be extended to a multiscale distribution akin to fractal analysis [83]. The report in [83] presents spectral analysis of multiscale Renyi entropy measures. Multiscale Renyi entropy has further information with regard to the common mean and variance measures, and should be applied as potential early markers of arrhythmia risk [83]. In [83–85] Renyi entropy is reported for the detection of cardiac autonomic neuropathy.

Using HRV analysis and principal component analysis (PCA), a CAN detection algorithm is reported in [86]. It employs multi-dimensional HRV data and two of the most significant components. It was verified by using data of 11 patients with definite CAN and 71 subjects without CAN and achieved a classification accuracy of 87%.

Recent work in CAN detection presents a predictive accuracy of 99.57%. It is based on automated iterative multiplier ensembles (AIME). However, the method includes blood biochemistry [87].

We provided a solution for detecting CAN in a portable or wearable healthcare device. A real-time processing architecture for CAN was realized that is suitable for battery powered or self-powered wearable devices. The proposed architecture is energy efficient, operating at an ultra-low power consumption.

In order to enable on-chip CAN classification, ECG processor needs to extract RR intervals and QT intervals. Several QRS detection and ECG feature extraction techniques have been reported in literature. Existing full ECG feature extraction algorithms that are implemented in Matlab include wavelet transform [27, 29] and low-pass differentiation [88]. Others have reported ECG feature extraction based on filtering and wavelet transform that are implemented in embedded systems [89]. The most common implemented techniques on-chip full feature extraction are based on wavelet transform [9, 13]. In addition, ultra-low power state-of-the-art ECG processors were proposed [13, 90] capable of detecting QRS complexes. Some systems have embedded a general-purpose CPU core [91], while others are custom accelerators [13, 90]. In [10] QRS detection is achieved by applying a multiscale Haar wavelet transform and maximum modulus pair recognition, whereas in [73] a quadratic spline wavelet transform is used. The system reported in [8] does QRS detection followed by classification.

The rest of the chapter is organized as follows. Section 6.2 reviews CAN detection algorithms and Sect. 6.3 presents an overview of the proposed full system. In Sect. 6.4 detailed description of the implemented architecture is provided. Power optimization is presented in Sect. 6.6. Section 6.5 gives the implementation results and discussion. In Sect. 6.7 an improved architecture for VA prediction is presented followed by summary in Sect. 6.8.

6.2 CAN Detection Algorithms

Non-invasive methods that are independent of patient cooperation are preferable in the diagnosis of CAN but still require further research to understand their sensitivity and specificity in risk assessment of CAN. Heart rate variability analysis is the most commonly used method [92]. Variation in HRV is regarded as one of the early signs of CAN [93, 94]. However, conventionally used time and frequency domain parameters of HRV are not always suitable for analysis because of the non-stationarity characteristics of the ECG recordings, the influence of respiration, and the presence of nonlinear phenomena in the physiological signal [95]. Several CAN detection algorithms either time or frequency domain have been reported in literature [49, 84, 96]. No clear evidence has been found to date of an alteration of sympathovagal balance attributed to the severity of CAN. Reductions in the conventional HRV parameters were reported as evidence of vagal efferent activity in diabetic patients [93]. However, alterations of sympathovagal balance with the severity of CAN were not detected. Here we summarize some of the reported techniques.

6.2.1 Tone-Entropy Technique

The technique presented in [96] demonstrates applying tone and entropy (T-E) in classifying CAN. Figure 6.1 shows the tone entropy space for three categories of CAN. The tone entropy represents individual and average value for normal (N), early (eCAN+), and definite (dCAN+) groups. Entropy decreased with the severity of CAN, however, tone increased with severity of CAN. Accordingly, the T-E method is appropriate for detecting the presence of CAN.

6.2.2 Time Domain RR-Based Methods

In [96], time domain RR heart rate variability parameters such as the mean RR, the standard deviation of normal RR data (SDNN), and the root mean square of successive difference (RMSSD) of RR data are demonstrated as indicators of CAN severity. These parameters decrease with severity of CAN, especially SDNN and RMSSD, which decreased significantly. However, mean RR was not significantly different between eCAN+ and dCAN+.

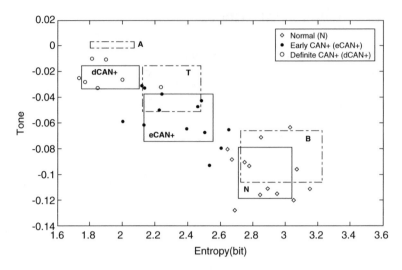

Fig. 6.1 Evaluated tone and entropy in T-E space [96]

6.2.3   QTVI-Based Methods

CAN detection based on QT variability index (QTVI) was proposed in [49] as a measure of QT and RR variability [97]. It was used to detect dilated cardiomyopathy (DCM) which is a cardiac arrhythmia associated with a high incidence of malignant ventricular arrhythmia and sudden death. The expression of QTVI is given in Eq. (6.1). QTVI is evaluated as the logarithm of the ratio of normalized QT variance to the heart rate variance. The mean and variance of the RR intervals as well as the mean and variance of the QT intervals are computed from the time series of heart rate (HR) and QT intervals [49].

$$QTVI = \log_{10} \left[\frac{\frac{QT_v}{QT_{m^2}}}{\frac{HR_v}{HR_{m^2}}} \right] \tag{6.1}$$

6.2.4   Renyi Entropy-Based Method

Determining the Renyi entropy for CAN detection and classification was proposed in [83–85]. Renyi entropy is defined as:

$$H(\alpha) = \frac{1}{1 - \alpha} \log_2 \left(\sum_{i=1}^{n} p_i^{\alpha} \right) \tag{6.2}$$

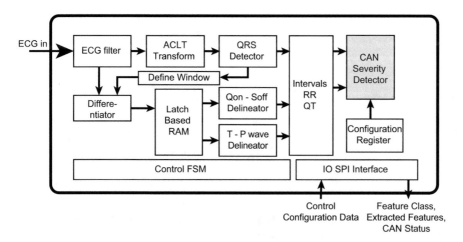

Fig. 6.2 ECG processing architecture with CAN detector

6.3 Proposed System Architecture

The proposed system architecture for ECG feature extraction and CAN classification is depicted in Fig. 6.2. It consists of two main parts which are the ECG feature extraction and CAN detector. In the first stage, ECG features are extracted that are utilized for the classification of CAN. These features are the main characteristic points of an ECG wave which are P-wave, QRS-complex, and T-wave. The peak, onset, and offset of these waves are also determined. In the second stage the extracted features are used to evaluate the QT and RR intervals, which are applied in detecting CAN severity.

The signal processing part is accomplished using an optimized algorithm that utilizes real-time and adaptive techniques for the detection and delineation of the P-QRS-T features of an ECG wave. These adaptive techniques are robust across ECG morphologies with high sensitivity and precision. Figure 6.2 shows the detailed block diagram of the system. Absolute value curve length transform (A-CLT) is used for detecting the QRS peaks, whereas QRS wave limits along with TP waves of the ECG are detected using low-pass differentiation. Memory is required to store samples to enable a backward search for relevant features. However, the required system memory is only 2 kB. Latch-based RAM was chosen in order to enable the system to operate at a low-supply voltage of 0.6 V. The latch-based RAM saved 60% in power-area product relative to an SRAM-based memory and more than 90% relative to a flip-flop-based RAM (Fig. 6.3). The power-area product is chosen as a metric for comparison because it gives an indication for the required resources and the power consumption.

The goal of our architecture is to provide real-time classification of CAN in an ultra-low power integrated biomedical system-on-chip (SoC). The CAN classification architecture is the hardware implementation of the algorithm described in [49]

Fig. 6.3 RAM type selection

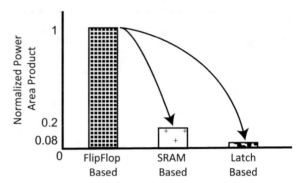

and [96]. As described in [49], classification of CAN is enabled by evaluating the QT variability index which is as shown in Eq. (6.1). Methods described in [96] include time series and frequency domain analysis of HRV. This chapter focuses on hardware realization of HRV time domain analysis with respect to classification of CAN severity.

The main features of the ECG are contained in the frequency range between 0.5 and 50 Hz. The sampling frequency of our proposed architectures is set to 250 Hz. For the current experiments the ECG was oversampled above Nyquist rate to enhance detection accuracy. Digitized ECG samples with 12-bits are directly filtered and the filter is pipelined operating at a clock rate of 250 Hz.

This section describes the utilized full ECG feature extraction starting with QRS peak detection followed by the full delineation process.

6.3.1 Proposed QRS Peak Detector

Figure 6.4 shows the proposed ACLT architecture for detecting the QRS complex. It is an architecture to realize the algorithm formulated in Eq. (6.3). It performs transformation, followed by QRS peak detection using an adaptive threshold technique. The transformation is completed by using derivative, absolute value, and integration all combined into one realization of the ACLT. The transformation distinctively enhances QRS complex even for noisy ECG signals that are corrupted with baseline wander (Fig. 6.5). Its unique inherent behavior removes the need for additional complicated circuits for high-pass or low-pass filters. All the computations for the transformation are performed using addition and shifting. Moreover, a comparison is required for detecting QRS peaks using thresholds. There is no need for multiplication, division, or applying square root manipulations. Hence the ACLT hardware implementation requires only adders, shifters, and comparators. These components are less hardware intensive relative to multipliers, dividers, and square root functions. For instance, if we compare an N-bit multiplier with an N-bit adder, an N-bit multiplier would require N times N-bit adders. Division and square

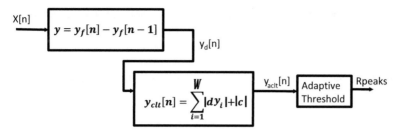

Fig. 6.4 Implemented ACLT

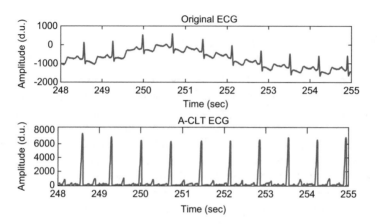

Fig. 6.5 ACLT for MIT-BIH record 112

root are much more complicated than addition or shifting.

$$L(\omega, i) = \sum_{i-\omega}^{i} \left| c^2 + |4 \times \Delta y_i| \right| \tag{6.3}$$

The integration over a window in the architecture is pipelined. Pipelining enables the signal to be transformed directly whenever there is a new ECG sample. Accordingly, the required clock frequency for the architecture is equal to the sampling frequency of the incoming ECG signal. The sampling frequency of the system is 250 Hz, accordingly, this is the lowest operating frequency possible that allows to process each acquired sample. Such a low operating frequency reduces the dynamic power dissipation. Depending on the proposed architecture duty cycling would not provide an advantage since the design is operating at the sampling rate of the incoming ECG signal. Buffering the ECG signal and then processing at higher frequency would require buffers (SRAM) which add more leakage power to the design power consumption.

QRS detection is achieved by applying a threshold on the transformed signal (Fig. 6.4). Obtaining an optimum threshold for wide morphological variant ECG signals is crucial for QRS detection. In the proposed architecture, the threshold is adaptive with respect to the detected QRS peaks. It is defined as in Eq. (6.4), where it is evaluated by accumulating R_{peak} values for eight beats and dividing the accumulated result by 16.

$$Th_{updated} = \frac{\sum_{i-8}^{i} Rpeak_i}{16} \tag{6.4}$$

6.3.2 ECG Feature Delineation

ECG feature delineation is finding the T-wave, P-wave (onset, peak, and offset) as well as QRS wave (onset and offset). Accurate delineation of such points in automated ECG processing systems is necessary for detecting cardiac rhythm abnormalities. This step follows the QRS peak detection. The flow chart of the feature extraction process is shown in Fig. 6.6. The full ECG feature extraction is achieved using differentiation and maximum modulus (MMP) pair recognition. In order to remove high frequency noise the differentiated signal is filtered before the feature delineation. This is illustrated in Fig. 6.6, waveforms 2 and 3. Adaptive windows are defined as probable location of these features, and the delineation process for each feature is described below.

6.3.2.1 QRS$_{on}$ and QRS$_{off}$ Detector

QRS$_{on}$ and QRS$_{off}$ are characterized by the minimum point or minimum slope before and after the R_{peak}, respectively. Differentiation enables identification of the minimum point or minimum slope. The normal width of the QRS complex ranges from 80 to 120 ms. In the proposed architecture a fixed offset of 40–120 ms relative to R_{peak} is used as the probable locations of QRS$_{on}$ and QRS$_{off}$.

6.3.2.2 Proposed T and P Wave Detector

Since the T-wave and P-wave are characterized by curved features and peaks, differentiation enables identification of the maximum slope and peaks. The window size for the P-wave ranges from $0.375 \times RR_{interval}$ up to $0.125 \times RR_{interval}$ before the R_{peak} location as illustrated in Fig. 6.7. Similarly the window size for the T-wave ranges from $0.065 \times RR_{interval}$ up to $0.65 \times RR_{interval}$ after the R_{peak} location (Fig. 6.7).

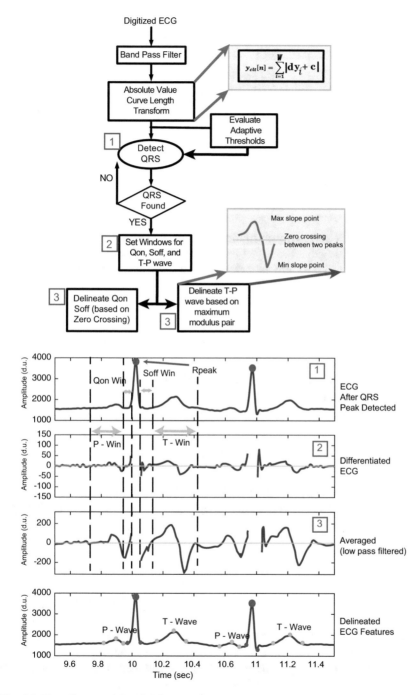

Fig. 6.6 Flow diagram of the ECG feature extraction

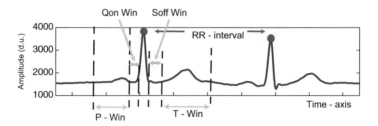

Fig. 6.7 Illustration of windows

6.3.3 ECG Intervals

In detecting CAN severity, the required features are the RR and the QT intervals [49]. Hence it is sufficient to detect the QRS peak, QRS onset, and the end of the T-wave. However, the feature extraction engine delineates all the ECG features as they are a part of the whole SoC.

6.4 Proposed CAN Severity Detector Architecture

In this section, we present the four CAN detector architectures, where the first three are implemented in the fabricated SoC.

6.4.1 QTVI-Based CAN Detection

The expression for QTVI is given in Eq. (6.1) [49] for classifying CAN severity. In order to ease hardware resources, the logarithm in Eq. (6.1) was removed as in Eq. (6.5). This gives us a linear expression for QTVI.

$$\text{QTVI} = \frac{\frac{\text{QT}v}{\text{QT}m^2}}{\frac{\text{HR}v}{\text{HR}m^2}} \tag{6.5}$$

Figure 6.8 demonstrates the architecture for evaluating the QTVI. As we can see from the figure the required operations are finding the mean and variance of the RR and QT intervals. Finding the mean requires accumulating RR or QT and dividing by the number of entries. Similarly the variance requires accumulating RR or QT and also accumulating the square of RR or QT, where the variance is evaluated as difference of expectations of variable X^2 and square of expectation of variable X. Variance of RR or QT is then divided by the respective mean squared. To perform all these operations, the main blocks are summation, multiplication, and division. 1024 RR and QT values were used so that the division for finding mean or variance could be implemented using shift operation.

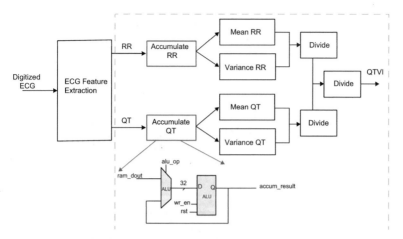

Fig. 6.8   QTVI evaluation architecture

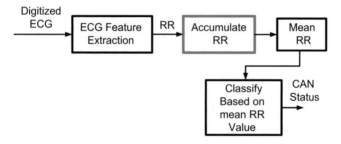

Fig. 6.9   Mean RR-based CAN classifying architecture

6.4.2   Mean RR-Based CAN Detection

The mean of RR intervals can be used to detect CAN severity as described in [96]. It is directly evaluated by accumulating extracted RR intervals from the ECG and dividing by the number of samples. To ease hardware implementation, 1024 points are taken to determine the mean. In [96] the mean for the three classes of CAN is presented. The same analysis is used in the proposed architecture to identify CAN severity. Figure 6.9 shows the architecture for detecting CAN based on mean RR.

6.4.3   RMSSD-Based CAN Detection

Root mean square value of standard difference (RMSSD) is another parameter that is applicable for CAN detection [96]. It is evaluated by calculating the RMS value of successive differences of the heart rate as shown in Fig. 6.10.

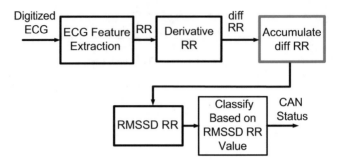

Fig. 6.10 RMSSD RR-based CAN classifying architecture

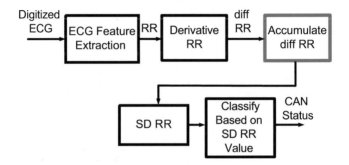

Fig. 6.11 SDRR-based CAN classifying architecture

6.4.4 SD-Based CAN Detection

Standard deviation (SD) is also used in CAN detection [96]. It is evaluated by calculating the standard deviation of the heart rate. Figure 6.11 shows the architecture for detecting CAN based on SD RR.

6.5 Results and Discussion

6.5.1 QRS Detection Results

Figure 6.12 shows the extracted QRS detection results, in which the extracted R_{peaks} are annotated on top of the ECG wave. Figure 6.12 demonstrates the matching between Matlab results and chip measurement results for the extracted RR intervals. The extraction results match with an error of less than 1%. Table 6.1 presents the performance comparison of the QRS detector in terms of sensitivity ($Se = TP/(TP + FN)$) and predictivity ($Pre = TP/(TP + FP)$). The ECG data from the MIT-BIH database [98] was used for validation and comparison. The

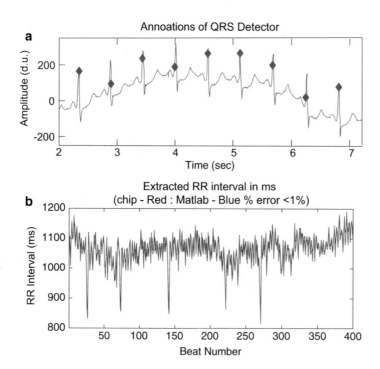

Fig. 6.12 (**a**) Annotation on extracted QRS peaks. (**b**) Extracted RR intervals in comparison with Matlab results

Table 6.1 Sensitivity and positive predictivity of QRS complex detectors (MIT-BIH)

Technique	Se	P+
[13]	99.29%	NA%
[9]	99.9%	99.9%
[10]	99.60%	99.77%
[73]	99.31%	99.70%
[64]	99.54%	99.74%
Proposed ACLT	99.37%	99.38%

proposed architecture achieves a sensitivity and predictivity greater than 99% which is acceptable for wearable healthcare devices. These values are also comparable to the reported systems as shown in Table 6.1.

6.5.2   ECG Feature Delineation Results

As described in section IIB, the purpose of the full feature extraction is to extract all the characteristic points of the ECG wave. From that the ECG intervals that are necessary for cardiac rhythm abnormality detection are evaluated. Figure 6.13

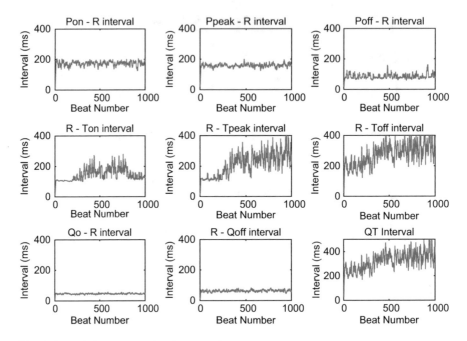

Fig. 6.13 ECG feature extraction demonstration

demonstrates the ECG intervals extracted for the record 103 from MIT-BIH database [98]. All extracted intervals lie within the acceptable range of the intervals. Normal ECG intervals are $P_{on} - R$ interval 0.12–0.2 s, QRS width 0.08–0.12 s, and QT interval 0.35–0.43 s.

6.5.3   CAN Detection Results

The CAN detection methods were validated using CAN database obtained from the Diabetes Complications Research Institute (DiScRi) [99]. The database contains ECG records for 223 patients attending the clinic between January and December 2008. CARTs were completed on all the participants who were categorized into three groups: without CAN (CAN−), early CAN (eCAN+), and definite or severe CAN (dCAN+) [81]. The CAN data was Lead II ECG over a 20 min recording which was sampled at 400 Hz. Since the implemented architecture works at 250 Hz, the data was resampled to 250 Hz. Eleven CAN−, 11 eCAN+, and 5 dCAN+ were used for validation of the realized CAN detection system. In the first case, these three types of CAN were analyzed based on RR interval variation. These include mean of RR, RMSSD, and SD of RR intervals as described in Sect. 6.3. The fourth technique was based on the combined variability of QT and RR, through evaluating QTVI as

Fig. 6.14 RR time series for
thee categories of CAN

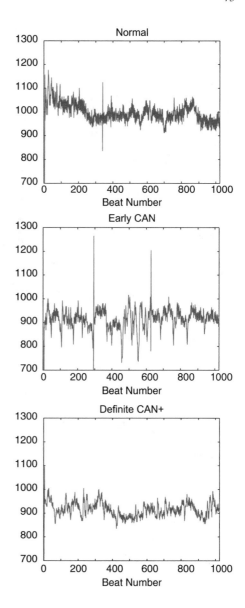

in Sect. 6.3.1. QTVI, mean of RR, and RMSSD architecture are implemented on-chip, however, the SD is evaluated in Matlab as a complementary analysis, where the RR intervals extracted from the chip are used in the evaluation of SD.

Figure 6.14 illustrates the RR time series for the three categories of CAN. Variability analysis of these intervals using mean, RMSSD, and SD could serve as indicators for CAN severity as illustrated in Fig. 6.15. Though the mean of the RR intervals varies with CAN severity, its variation is not statistically significant.

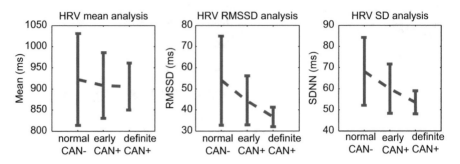

Fig. 6.15 Time domain HRV parameters

Fig. 6.16 Error bar plot for
extracted QTc

Fig. 6.17 Linear QTVI for
CAN classification

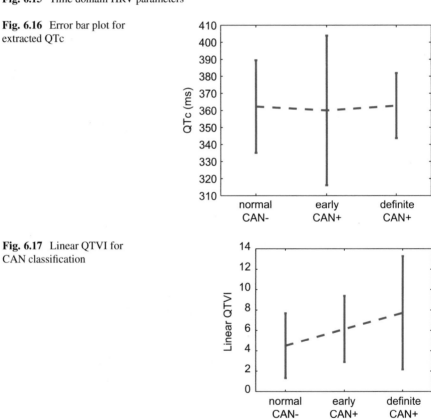

However the RMSSD was observed to decrease with CAN severity, especially CAN− and dCAN+ differ with a p-value less than 0.05. Similarly, SD varies for the different categories of CAN for the investigated dataset.

Figures 6.16 and 6.17 reveal QTc and linear QTVI error bar plots for CAN−, eCAN+ and dCAN+. QTc alone does not reveal significant differences for these three categories of CAN. However, the linear QTVI increased with severity of CAN.

This indicates that linear QTVI is a good candidate for detecting CAN severity on ECG processing SoCs considering its ultra-low power implementation.

6.6 Chip Implementation and Power Optimization

The proposed SoC was implemented and fabricated using the 65 nm CMOS process and standard cell library. These standard cells had different flavors such as RVT, HVT, and LVT. High voltage threshold (HVT) standard cells were used to synthesize the system, so as to reduce the leakage. Overall power dissipation was reduced by operating at near threshold. Operating voltage of the system was scaled down to 0.6 V, which is near to the threshold voltage of 65 nm CMOS technology. The standard cells library on which the system was fabricated can operate down to 0.6 V, however, the SRAM had a minimum operating voltage of 0.9 V. Hence, to power the system at 0.6 V a latch-based memory was used. Table 6.2 lists the power consumption at supply voltages 0.6–1.1 V. The power consumption is dominated by leakage in all the cases. Probable ways of reducing the energy dissipation is to enable clock gating or power gating. However, the system does not benefit from clock gating since the dynamic power is linearly related to the operating clock frequency. If we double the clock frequency, the dynamic power would be doubled. The leakage energy over time remains the same. If we want to enable power gating by operating at higher frequency, extra circuitry would be required to save the state of the system. The system memory will be on all the time, which is where the dominant leakage is coming from.

The chip micrograph along with the measured performance summary and measurement setup is provided in Fig. 6.18. The ECG process occupied an area of 0.108 mm^2 and consumed 75 nW when operating at 0.6 V and 250 Hz. ECG function for feature extraction and CAN classification was validated by injecting digitized ECG data through an FPGA. The data is stored in an SD card and the chip communicates with the FPGA through an SPI interface. The measurement results are captured using a logic analyzer as depicted in Fig. 6.18. Power consumption is measured by power supply with built-in DC power analyzer.

In Table 6.3 the comparison with the published work is presented. The realized CAN detection architecture does full ECG feature extraction and CAN detection. Its power consumption is less than the related systems [8–10, 13, 90]. Though the SoCs reported in [8, 13] include instrumentation amplifier and ADC, their respective digital power is dominant. The CAN detector also occupied relatively the lowest area.

Fig. 6.18 Die photo and
specification summary

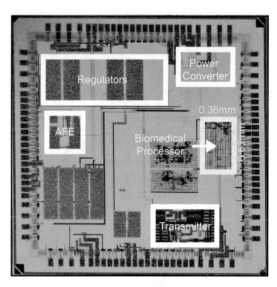

Supply Voltage	Operating Frequency	Area	Memory	Power Consumption
0.6V	250Hz	0.108 mm²	2 kb	75nW

Table 6.2 Leakage power of the SoC

Supply voltage (V)	0.6	0.7	0.8	0.9	1.0	1.1
Leakage power (nW)	55.8	94.5	174.4	313.2	537	913
Dynamic power (nW)	19.2	27.3	25.6	28.8	53	46.75
Total power (nW)	75	121.8	200	342	590	959.75

Table 6.3 Performance comparison with published work

	[10]	[8]	[13]	[9]	[90]	This work
Technology (nm)	180	180	180	350	65	65
Area mm^2	0.484	2.47	NA	1.2	0.243	0.108
Oper. freq. (kHz)	NA	0.12	500	1	7.5	250
Supply voltage	1.0	1.2	0.5	3	0.6	0.6
AFE	NA	IA ADC	IA ADC	SAR ADC	NA	NA
ECG features	QRS	QRS	P-QRS-T	QRS	P-QRS-T	P-QRS-T
Power (μW)	0.410	5.97	0.457	13.6	0.642	0.075
Energy per cycle (nJ)	NA	49.75	0.914	13.6	0.0856	0.3

Fig. 6.19 Schematic representation of the proposed ventricular arrhythmia prediction system

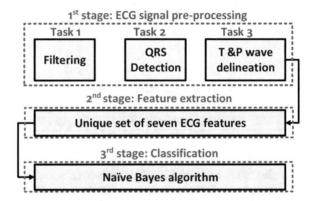

6.7   VTVF Classifier Architecture

The baseline VA processor presented in [12] consists of three main stages: ECG pre-processing, feature extraction, and classification as shown in Fig. 6.19. In this section we present the improved architecture of the VA processor [100]. In the ECG pre-processing stage all the ECG wave features such as QRS complex, T-wave, and P-wave are extracted. Prior to ECG delineation, filtering is performed because ECG could be corrupted by baseline drift, power-line interference, and high frequency noise. After filtering, QRS detection is performed based on the Pan and Tompkins technique [24]. Along with the QRS peak detection, the QRS onset and offset are also delineated. Finally, T and P waves are delineated, and the corresponding fiducial points (P onset, P peak, P offset, T onset, T peak, and T offset) are extracted.

In this work three major modifications were carried out to lower the power consumption: (1) Elimination of the SRAM block in QRS detection, and thus, reducing RAM requirements from 8 kB down to 4 kB. (2) Decreasing the operating frequency to 250 Hz equal to the ECG sampling frequency. (3) The use of high threshold voltage (HVT) cells to reduce the leakage power. The details of the optimization are discussed in the following subsection.

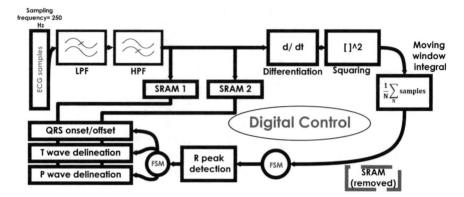

Fig. 6.20 Block diagram of pre-processing stage, which contains filtering, QRS detection, and P and T wave delineation. Only 4 kB RAM is required in the proposed architecture, as compared to baseline 8 kB. The proposed architecture does not require an SRAM to store the transformed samples

6.7.1   ECG Pre-processing

In the baseline architecture [12], QRS detection was performed on samples that were stored in SRAM. However, in the updated architecture, QRS peaks are detected directly from the new samples, eliminating the need for storing transformed samples. Such a seemingly minor improvement actually leads to major reductions in area and power consumption. As illustrated in Fig. 6.20, the output of the moving integral is directly compared to the threshold in the R-peak detection module. Such an optimization enables the system to reach real-time operation even when with clock frequency equal to the sampling frequency. Other features such as the onset/offset of QRS, and onset/peak/offset of T and P waves are computed using filtered samples which are stored in the memory. Two SRAM blocks are used in the architecture, which are used alternatively for storing new samples and reading/processing simultaneously. Whenever the delineation is done from SRAM 1 (Fig. 6.20), new filtered signals are stored in the SRAM 2, and vice versa. Note that the delineation and switching functions (read/write) of the two SRAMs are performed when a new QRS peak is detected.

6.7.2   Feature Extraction Stage

This stage extracts the features that are used by the classifier. We start by determining the unique set of ECG features that best capture the characteristics of ventricular arrhythmia. The features include: RR, PQ, QP, RT, TR, PS, and SP intervals [12]. Figure 6.21 shows these intervals on a representative ECG record. It

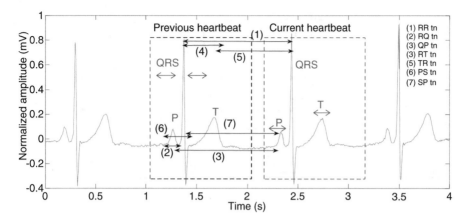

Fig. 6.21 An annotated ECG waveform highlighting ECG wave components and the features extracted for the ECG classifier

is worth mentioning that the features are extracted from two consecutive heartbeats, unlike other methods which process each heartbeat independently. Utilizing two consecutive heartbeats give higher accuracy in predicting VA.

6.7.3 Classification Stage

The objective of the classifier is to predict the presence of VA based on the features mentioned above. Since our system aims at ultra-lower power implementation, only a classifier architecture that requires minimum hardware resources can be afforded. Having this in mind, the Naive Bayes classifier was chosen. The Naive Bayes classifier is a linear classifier and easy to build with no complicated iterative parameter estimation. It assumes naive and strong independent distributions between the feature vectors, and this assumption was met since all the extracted ECG features were analyzed and assessed independently from the beginning [12].

Figure 6.21 shows the six intervals on the ECG record [12]. These features were thoroughly studied to be the best indicator for VT/VF condition and for the databases of VA recordings from American Heart Association (AHA). Naive Bayes classifier was utilized for classification that can predict normal and VA situations. Mainly, the high differentiation quality of the signals has enabled the successful use of a linear classifier for the decision-making process.

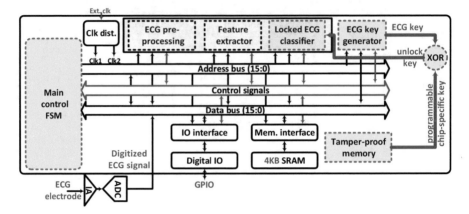

Fig. 6.22 Architecture of the proposed secure ventricular arrhythmia prediction processor. The blocks relevant to security are highlighted in orange color with dashed borders

6.7.4   Secure VA Prediction Architecture

The proposed improved architecture for the VA predictor is integrated into secure processor IoT sensing platform as shown in Fig. 6.22 [100]. The proposed solution is able to predict the onset of VA upto 3 h in advance with 86% accuracy where the classification accuracy is computed as in Eq. (6.6). In Eq. (6.6), TN refers to the number of true negative detections, TP refers to the number of true positive detections, FP to number of false-positive detection, and FN to number of false negative detections. Moreover, the proposed architecture is designed using an application specific integrated circuits design flow in 65 nm LPe CMOS technology; the power it consumes is 62.2% less than that of the state-of-the-art approaches [12], while occupying 16.0% smaller area. The proposed processor makes use of ECG signals to extract a chip-specific ECG key that enables protection of communication channel. By integrating the ECG key with an existing design-for-trust solution, the proposed platform offers protection also at the hardware level, thwarting hardware security threats, such as reverse engineering and counterfeiting. Through efficient sharing of on-chip resources, the overhead of the multi-layered security infrastructure is kept at 9.5% for area and 0.7% for power with no impact on the speed of the design.

$$\text{ACC} = \frac{TP + TN}{TP + TN + FP + FN}. \tag{6.6}$$

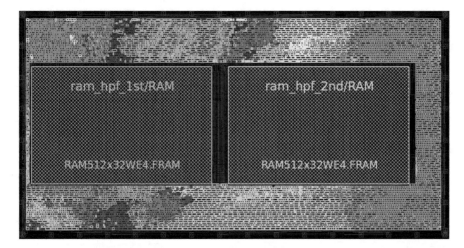

Fig. 6.23 Layout of proposed VA processor

6.7.5 ASIC Design

The VA processor was designed using Verilog-HDL and the layout was generated using the standard physical design flow; the 65 nm LPe library was used. The design was synthesized by using the Synopsys design compiler, and then floor-planned, placed, and routed by using Synopsys ICC compiler. The provided results are based on post-layout power and area analysis. Since the design is purely digital, it is natural to expect a very high correlation (within 5% level as accepted in the chip-design industry) between the post-silicon results and the post-layout design.

Figure 6.23 shows the floorplan of the design. The placed and routed VA processor footprint dimension was 430 μm by 218 μm with a total utilization of 85%. Table 6.4 presents the design results. The design occupied a total core area of 0.0941 mm^2 and consumed a total power of 1.052 μW at 250 Hz frequency. The implementation reported in [12] uses 8 kB SRAMs, 4 kB for the QRS detector, and 4 kB for detecting the other features. In our improved architecture depicted in Fig. 6.22, only 4 kB SRAM is needed for detecting the QRS onset/offset, T-wave, and P-wave. The QRS peaks are directly detected from the output of the moving average signal. Such scheme reduces the total power by 62.2% and area by 16.0%. Moreover this new architecture operates at the same frequency as the sampling frequency. This is the minimum frequency at which ECG processors could operate in order to fulfill the real-time requirement. Operating at low frequency also reduces the dynamic power.

The comparison of the proposed VA processor with the existing ECG processors is presented in Table 6.5. The proposed architecture performs prediction based on multiple ECG features extracted from the P-wave, the T-wave, and the QRS complex; on the other hand, the previous processors [8, 52, 101] perform detection

Table 6.4 Secure VA
processor design metrics

Parameter	Figure
Total area	$0.0941\,\text{mm}^2$
Dynamic power	$9.71e^{-3}\,\mu\text{W}$
Leakage power	$1.042\,\mu\text{W}$
Total power	$1.052\,\mu\text{W}$
Utilization	85%

based on RR interval only. Although the power consumption reported in [52] is lower than our proposed processor, [52] is dominated by analog blocks and it only performs arrhythmia classification with no prediction. The proposed VA processor has lower power consumption and area compared to [101] and [8].

6.8   Summary

An ultra-low power ECG processing architecture that comprises ECG feature extraction and CAN classification was presented. The proposed architecture operates at an ultra-low power dissipation through algorithmic and architectural optimization. An absolute value CLT detects QRS peaks with sensitivity of 99.37% and predictivity of 99.38%. Full ECG feature extraction is enabled through differentiation, windowing, and threshold. System memory was implemented using latch-based memory, which enables the system to operate down to 0.6 V. The system was fabricated using 65 nm low power technology and consumed 75 nW only when operating at a frequency of 250 Hz from a supply voltage of 0.6 V. The CAN severity utilizes the extracted features. Twenty seven records were utilized to verify the detection processor, and the QT variability index (QTVI) and RR interval variability analysis were validated as indicators for differentiating control, early and definite CAN.

Moreover, an improved architecture for VA prediction was presented in this chapter. The improved VA prediction architecture achieved a reduction in the required area by 16.0% and reduction in power consumption by 62.2%.

Table 6.5 Comparison of (secure) VA processor hardware with previously implemented designs

Reference	Technology (μm)	Area (mm^2)	Power (μW)	Supply voltage (V)	Frequency (KHz)	Detection accuracy (%)	Prediction accuracy (%)	Security
[101]	0.180	2.250	1.260	0.7	1	NA	NA	Insecure
[8]	0.180	2.465	5.967	1.2	0.12	97.25	NA	Insecure
[52]	0.065	3.321	0.095	0.4	10	NA	NA	Insecure
Baseline VA [12]	0.065	0.112	2.780	1.2	10	NA	86	Insecure
Proposed VA	0.065	0.094	1.052	1.2	0.25	NA	86	Insecure
Proposed secure VA	0.065	0.103	1.059	1.2	0.25	NA	86	Secure

Bibliography

1. WHO, Life expectancy at birth (2015)
2. WHO, The top 10 causes of death, Jan 2017
3. A. Klinefelter, N.E. Roberts, Y. Shakhsheer, P. Gonzalez, A. Shrivastava, A. Roy, K. Craig, M. Faisal, J. Boley, S. Oh, Y. Zhang, D. Akella, D.D. Wentzloff, B.H. Calhoun, A 6.45μW self-powered IoT SoC with integrated energy-harvesting power management and ULP asymmetric radios, in *2015 IEEE International Solid- State Circuits Conference - (ISSCC)*, Feb 2015, pp. 1–3
4. J. Kwong, A.P. Chandrakasan, An energy-efficient biomedical signal processing platform. IEEE J. Solid State Circuits **46**(7), 1742–1753 (2011)
5. M. Konijnenburg, Y. Cho, M. Ashouei, T. Gemmeke, C. Kim, J. Hulzink, J. Stuyt, M. Jung, J. Huisken, S. Ryu et al., Reliable and energy-efficient 1MHz 0.4 V dynamically reconfigurable SoC for ExG applications in 40nm LP CMOS, in *2013 IEEE International Solid-State Circuits Conference Digest of Technical Papers (ISSCC)* (IEEE, New York, 2013), pp. 430–431
6. Y. Zou, J. Han, S. Xuan, S. Huang, X. Weng, D. Fang, X. Zeng, An energy-efficient design for ECG recording and R-peak detection based on wavelet transform. IEEE Trans. Circuits Syst. Express Briefs **62**(2), 119–123 (2015)
7. S. Izumi, K. Yamashita, M. Nakano, S. Yoshimoto, T. Nakagawa, Y. Nakai, H. Kawaguchi, H. Kimura, K. Marumoto, T. Fuchikami, Y. Fujimori, H. Nakajima, T. Shiga, M. Yoshimoto, Normally off ECG SoC with non-volatile MCU and noise tolerant heartbeat detector. IEEE Trans. Biomed. Circuits Syst. **9**(5), 641–651 (2015)
8. S.Y. Lee, J.H. Hong, C.H. Hsieh, M.C. Liang, S.Y.C. Chien, K.H. Lin, Low-power wireless ECG acquisition and classification system for body sensor networks. IEEE J. Biomed. Health Inf. **19**(1), 236–246 (2015)
9. Y.-J. Min, H.-K. Kim, Y.-R. Kang, G.-S. Kim, J. Park, S.-W. Kim, Design of wavelet-based ECG detector for implantable cardiac pacemakers. IEEE Trans. Biomed. Circuits Syst. **7**(4), 426–436 (2013)
10. P. Li, X. Zhang, M. Liu, X. Hu, B. Pang, Z. Yao, H. Jiang, H. Chen, A 410-nW efficient QRS processor for mobile ECG monitoring in 0.18-μm CMOS. IEEE Trans. Biomed. Circuits Syst. **11**(6), 1356–1365 (2017)
11. R.A. Abdallah, N.R. Shanbhag, An energy-efficient ECG processor in 45-nm CMOS using statistical error compensation. IEEE J. Solid State Circuits **48**(11), 2882–2893 (2013)

© Springer International Publishing AG, part of Springer Nature 2019
T. Tekeste Habte et al., *Ultra Low Power ECG Processing System for IoT Devices*, Analog Circuits and Signal Processing,
https://doi.org/10.1007/978-3-319-97016-5

12. N. Bayasi, T. Tekeste, H. Saleh, B. Mohammad, A. Khandoker, M. Ismail, Low-power ECG-based processor for predicting ventricular arrhythmia. IEEE Trans. Very Large Scale Integr. VLSI Syst. **24**(5), 1962–1974 (2016)

13. X. Liu, J. Zhou, Y. Yang, B. Wang, J. Lan, C. Wang, J. Luo, W.L. Goh, T.T.-H. Kim, M. Je, A 457 nW near-threshold cognitive multi-functional ECG processor for long-term cardiac monitoring. IEEE J. Solid State Circuits **49**(11), 2422–2434 (2014)

14. F. Zhang, Y. Zhang, J. Silver, Y. Shakhsheer, M. Nagaraju, A. Klinefelter, J. Pandey, J. Boley, E. Carlson, A. Shrivastava et al., A batteryless 19μW MICS/ISM-band energy harvesting body area sensor node SoC, in *2012 IEEE International Solid-State Circuits Conference Digest of Technical Papers (ISSCC)* (IEEE, New York, 2012), pp. 298–300

15. S.M.R. Islam, D. Kwak, M.H. Kabir, M. Hossain, K.S. Kwak, The internet of things for health care: a comprehensive survey. IEEE Access **3**, 678–708 (2015)

16. R.S.H. Istepanian, S. Hu, N.Y. Phiip, A. Sungoor, The potential of internet of m-health things "m-iot" for non-invasive glucose level sensing, in *2011 Annual International Conference of the IEEE Engineering in Medicine and Biology Society*, Aug 2011, pp. 5264–5266

17. Q. Li, C. Rajagopalan, G.D. Clifford, Ventricular fibrillation and tachycardia classification using a machine learning approach. IEEE Trans. Biomed. Eng. **61**(6), 1607–1613 (2014)

18. A. Dohr, R. Modre-Opsrian, M. Drobics, D. Hayn, G. Schreier, The internet of things for ambient assisted living, in *2010 Seventh International Conference on Information Technology: New Generations*, April 2010, pp. 804–809

19. H.A. Khattak, M. Ruta, E. Di Sciascio, D. Sciascio, Coap-based healthcare sensor networks: a survey, in *Proceedings of 2014 11th International Bhurban Conference on Applied Sciences Technology (IBCAST)* Islamabad, 14th–18th Jan 2014, pp. 499–503

20. L.P. Son, N.T.A. Thu, N.T. Kien, Design an IoT wrist-device for SpO2 measurement, in *2017 International Conference on Advanced Technologies for Communications (ATC)*, Oct 2017, pp. 144–149

21. WHO, Cardiovascular diseases (CVDs), Oct 2015

22. G.D. Clifford, F. Azuaje, P.E. McSharry, *Advanced Methods and Tools for ECG Data Analysis* (Artech House, Boston, 2006)

23. P. Kligfield, The centennial of the Einthoven electrocardiogram. J. Electrocardiogr. **35**(4), 123–129 (2002)

24. J. Pan, W.J. Tompkins, A real-time QRS detection algorithm. IEEE Trans. Biomed. Eng. **32**(3), 230–236 (1985)

25. P.S. Hamilton, W.J. Tompkins, Quantitative investigation of QRS detection rules using the MIT/BIH arrhythmia database. IEEE Trans. Biomed. Eng. **33**(12), 1157–1165 (1986)

26. M. Niknazar, B. Rivet, C. Jutten, Fetal ECG extraction by extended state Kalman filtering based on single-channel recordings. IEEE Trans. Biomed. Eng. **60**(5), 1345–1352 (2013)

27. J. Martinez, R. Almeida et al., A wavelet-based ECG delineator: evaluation on standard database. IEEE Trans. Biomed. Eng. **51**(4), 570–348 (2004)

28. M.W. Phyu, Y. Zheng, B. Zhao, L. Xin, Y.S. Wang, A real-time ECG QRS detection ASIC based on wavelet multiscale analysis, in *IEEE Asian Solid-State Circuits Conference* (IEEE, New York, 2009), pp. 293–296

29. E.B. Mazomenos, D. Biswas, A. Acharyya, T. Chen, K. Maharatna, J. Rosengarten, J. Morgan, N. Curzen, A low-complexity ECG feature extraction algorithm for mobile healthcare applications. IEEE J. Biomed. Health Inf. **17**(2), 459–469 (2013)

30. S. Banerjee, M. Mitra, Application of cross wavelet transform for ECG pattern analysis and classification. IEEE Trans. Instrum. Meas. **63**(2), 326–333 (2014)

31. I.S. Murthy, G.S. Prasad, Analysis of ECG from pole-zero models. IEEE Trans. Biomed. Eng. **39**(7), 741–751 (1992)

32. S. Pal, M. Mitra, Empirical mode decomposition based ECG enhancement and QRS detection. Comput. Biol. Med. **42**(1), 83–92 (2012)

33. R.J. Oweis, E.W. Abdulhay, Seizure classification in EEG signals utilizing Hilbert-Huang transform. Biomed. Eng. Online **10**, 38 (2011)

34. A. Khawaja, Automatic ECG analysis using principal component analysis and wavelet transformation. University-Verlag Karlsruhe (2006). https://doi.org/10.5445/KSP/1000006642
35. S.S. Mehta, S.C. Saxena, H.K. Verma, Recognition of P and T waves in electrocardiograms using fuzzy theory, in *Engineering in Medicine and Biology Society, IEEE 14th Conference of the Biomedical Engineering Society of India. An International Meeting, Proceedings of the First Regional Conference* (IEEE, New York, 1995), pp. 2–54
36. E. de Azevedo Botter, C.L. Nascimento, T. Yoneyama, A neural network with asymmetric basis functions for feature extraction of ECG P waves. IEEE Trans. Neural Netw. **12**(5), 1252–1255 (2001)
37. G.N. Golpayegani, A.H. Jafari, A novel approach in ECG beat recognition using adaptive neural fuzzy filter. J. Biomed. Sci. Eng. **2**, 80–85 (2009)
38. S.S. Mehta, N.S. Lingayat, Detection of QRS complexes in electrocardiogram using support vector machine. J. Med. Eng. Technol. **32**(3), 206–215 (2008)
39. M.R. Homaeinezhad, S.A. Atyabi, E. Tavakkoli, H.N. Toosi, A. Ghaffari, R. Ebrahimpour, ECG arrhythmia recognition via a Neuro-SVM–KNN hybrid classifier with virtual QRS image-based geometrical features. Expert Syst. Appl. **39**(2), 2047–2058 (2012)
40. R.N. Khushaba, S. Kodagoda, S. Lal, G. Dissanayake, Driver drowsiness classification using fuzzy wavelet-packet-based feature-extraction algorithm. IEEE Trans. Biomed. Eng. **58**(1), 121–131 (2011)
41. Y.N. Singh, P. Gupta, ECG to individual identification, in *2nd IEEE International Conference on Biometrics: Theory, Applications and Systems* (2008), pp. 1–8
42. K.F. Tan, K.L. Chan, K. Choi, Detection of the QRS complex, P wave and T wave in electrocardiogram, in *IEEE Conference on Medical Signal and Information Processing* (2000), pp. 41–47
43. M. Elgendi, M. Jonkman, F. De Boer, Recognition of T waves in ECG signals, in *IEEE Conference on Bioengineering* (2009), pp. 1–2
44. E.B. Mazomenos, T. Chen, A. Acharyya, A. Bhattacharya, J. Rosengarten, K. Maharatna, A time-domain morphology and gradient based algorithm for ECG feature extraction, in *2012 IEEE International Conference on Industrial Technology (ICIT)* (IEEE, New York, 2012), pp. 117–122
45. A.A. Cost, G.G. Cano, QRS detection based on hidden Markov modeling, in *Proceedings of the Annual International Conference of the IEEE Engineering in Engineering in Medicine and Biology Society, 1989. Images of the Twenty-First Century* (IEEE, New York, 1989), pp. 34–35
46. C. Lin, C. Mailhes, J.-Y. Tourneret, P- and T-wave delineation in ECG signals using a Bayesian approach and a partially collapsed Gibbs sampler. IEEE Trans. Biomed. Eng. **57**(12), 2840–2849 (2010)
47. S. Mallat, *A Wavelet Tour of Signal Processing*, 2nd edn. (Academic, New York, 1999)
48. Z. Chen, J. Luo, K. Lin, J. Wu, T. Zhu, X. Xiang, J. Meng, An energy-efficient ECG processor with weak-strong hybrid classifier for arrhythmia detection. IEEE Trans. Circuits Syst. Express Briefs (2017). https://doi.org/10.1109/TCSII.2017.2747596
49. A.H. Khandoker, M.H. Imam, J. Couderc, M. Palaniswami, H.F. Jelinek, QT variability index changes with severity of cardiovascular autonomic neuropathy. IEEE Trans. Inf. Technol. Biomed. **16**(5), 900–906 (2012)
50. D. Jeon, Y.-P. Chen, Y. Lee, Y. Kim, Z. Foo, G. Kruger, H. Oral, O. Berenfeld, Z. Zhang, D. Blaauw, D. Sylvester, An implantable 64nW ECG-monitoring mixed-signal SoC for arrhythmia diagnosis, in *2014 IEEE International Solid-State Circuits Conference Digest of Technical Papers (ISSCC)*, Feb 2014, pp. 416–417
51. M. Alioto, Ultra-low power VLSI circuit design demystified and explained: a tutorial. IEEE Trans. Circuits Syst. Regul. Pap. **59**(1), 3–29 (2012)
52. Y.P. Chen, D. Jeon, Y. Lee, Y. Kim, Z. Foo, I. Lee, N.B. Langhals, G. Kruger, H. Oral, O. Berenfeld, Z. Zhang, D. Blaauw, D. Sylvester, An injectable 64 nW ECG mixed-signal SoC in 65 nm for Arrhythmia monitoring. IEEE J. Solid State Circuits **50**(1), 375–390 (2015)

53. R. Almeida, J.P. Martinez, S. Olmos, A.P. Rocha, P. Laguna, Automatic delineation of T and P waves using a wavelet-based multiscale approach, in *Proceedings of the 1st International Congress on Computational Bioengineering* (2003), pp. 243–247
54. W. Zong, G.B. Moody, D. Jiang, A robust open-source algorithm to detect onset and duration of QRS complexes, in *Computers in Cardiology, 2003* (IEEE, New York, 2003), pp. 737–740
55. M. Keating, D. Flynn, R. Aitken, A. Gibbons, K. Shi, *Low Power Methodology Manual: For System-on-Chip Design* (Springer, Berlin, 2007)
56. TI, Ultra-low power comparison: MSP430 vs. microchip XLP tech brief (2009)
57. M. Alhawari, B. Mohammad, H. Saleh, M. Elnaggar, An efficient zero current switching control for L-based DC-DC converters in TEG applications. IEEE Trans. Circuits Syst. Express Briefs **64**, 294–298 (2016)
58. G.M. Friesen, T.C. Jannett, M.A. Jadallah, S.L. Yates, S.R. Quint, H.T. Nagle, A comparison of the noise sensitivity of nine QRS detection algorithms. IEEE Trans. Biomed. Eng. **37**(1), 85–98 (1990)
59. T. Tekeste, N. Bayasi, H. Saleh, A. Khandoker, B. Mohammad, M. Al-Qutayri, M. Ismail, Adaptive ECG interval extraction, in *2015 IEEE International Symposium on Circuits and Systems (ISCAS)* (IEEE, New York, 2015), pp. 998–1001
60. L. Catarinucci, D. de Donno, L. Mainetti, L. Palano, L. Patrono, M.L. Stefanizzi, L. Tarricone, An IoT-aware architecture for smart healthcare systems. IEEE Internet Things J. **2**(6), 515–526 (2015)
61. M. Hassanalieragh, A. Page, T. Soyata, G. Sharma, M. Aktas, G. Mateos, B. Kantarci, S. Andreescu, Health monitoring and management using internet-of-things (IoT) sensing with cloud-based processing: opportunities and challenges, in *2015 IEEE International Conference on Services Computing (SCC)* (IEEE, New York, 2015), pp. 285–292
62. S.Y. Ge, S.M. Chun, H.S. Kim, J.T. Park, Design and implementation of interoperable IoT healthcare system based on international standards, in *13th IEEE Annual Consumer Communications Networking Conference (CCNC)*, Jan 2016, pp. 119–124
63. F. Fernandez, G.C. Pallis, Opportunities and challenges of the internet of things for healthcare: systems engineering perspective, in *4th International Conference on Wireless Mobile Communication and Healthcare - Transforming Healthcare Through Innovations in Mobile and Wireless Technologies (MOBIHEALTH)*, Nov 2014, pp. 263–266
64. R. Gutiérrez-Rivas, J.J. García, W.P. Marnane, Á. Hernández, Novel real-time low-complexity QRS complex detector based on adaptive thresholding. IEEE Sensors J. **15**(10), 6036–6043 (2015)
65. C.J. Deepu, X. Zhang, W.-S. Liew, D.L.T. Wong, Y. Lian, An ECG-on-Chip with 535 nW/channel integrated lossless data compressor for wireless sensors. IEEE J. Solid State Circuits **49**(11), 2435–2448 (2014)
66. S.L. Chen, J.G. Wang, VLSI implementation of low-power cost-efficient lossless ECG encoder design for wireless healthcare monitoring application. Electron. Lett. **49**(2), 91–93 (2013)
67. T. Marisa, T. Niederhauser, A. Haeberlin, R.A. Wildhaber, R. Vogel, M. Jacomet, J. Goette, Bufferless compression of asynchronously sampled ECG signals in cubic Hermitian vector space. IEEE Trans. Biomed. Eng. **62**(12), 2878–2887 (2015)
68. S. Feizi, G. Angelopoulos, V.K. Goyal, M. Médard, Backward adaptation for power efficient sampling. IEEE Trans. Signal Process. **62**(16), 4327–4338 (2014)
69. A.M.R. Dixon, E.G. Allstot, D. Gangopadhyay, D.J. Allstot, Compressed sensing system considerations for ECG and EMG wireless biosensors. IEEE Trans. Biomed. Circuits Syst. **6**(2), 156–166 (2012)
70. TI, MSP430FR2433 mixed-signal microcontroller (2015)
71. ATMEL, ATMEL 8-BIT Microcontroller with 4/8/16/32Kbytes in-system programmable flash (2015)
72. M. Alhawari, D. Kilani, B. Mohammad, H. Saleh, M. Ismail, An efficient thermal energy harvesting and power management for μWatt wearable biochips, in *2016 IEEE International Symposium on Circuits and Systems (ISCAS)* (IEEE, New York, 2016), pp. 2258–2261

73. C.-I. Ieong, P.-I. Mak, C.-P. Lam, C. Dong, M.-I. Vai, P.-U. Mak, S.-H. Pun, F. Wan, R.P. Martins, A 0.83-QRS detection processor using quadratic spline wavelet transform for wireless ECG acquisition in 0.35-CMOS. IEEE Trans. Biomed. Circuits Syst. **6**(6), 586–595 (2012)

74. C.J. Deepu, C.-H. Heng, Y. Lian, A hybrid data compression scheme for power reduction in wireless sensors for IoT. IEEE Trans. Biomed. Circuits Syst. **11**(2), 245–254 (2016)

75. M. Ishijima, S.B. Shin, G.H. Hostetter, J. Sklansky, Scan-along polygonal approximation for data compression of electrocardiograms. IEEE Trans. Biomed. Eng. **BME-30**(11), 723–729 (1983)

76. E. Tsimbalo, X. Fafoutis, R.J. Piechocki, CRC error correction in IoT applications. IEEE Trans. Ind. Inf. **13**(1), 361–369 (2017)

77. A.L. Goldberger, L.A.N. Amaral, L. Glass, J.M. Hausdorff, P.C. Ivanov, R.G. Mark, J.E. Mietus, G.B. Moody, C.-K. Peng, H.E. Stanley, PhysioBank, PhysioToolkit, and PhysioNet: components of a new research resource for complex physiologic signals. Circulation **101**(23), e215–e220 (2000). Circulation Electronic Pages: http://circ.ahajournals.org/cgi/content/full/101/23/e215. PMID:1085218; https://doi.org/10.1161/01.CIR.101.23.e215

78. P. Laguna, R.G. Mark, A. Goldberg, G.B. Moody, A database for evaluation of algorithms for measurement of QT and other waveform intervals in the ECG, in *Computers in Cardiology* (IEEE, New York, 1997), pp. 673–676

79. G.B. Moody, R.G. Mark, The impact of the MIT-BIH arrhythmia database. IEEE Eng. Med. Biol. Mag. **20**(3), 45–50 (2001)

80. A.I. Vinik, R.E. Maser, B.D. Mitchell, R. Freeman, Diabetic autonomic neuropathy. Diabetes Care **26**(5), 1553–1579 (2003)

81. D.J. Ewing, C.N. Martyn, R.J. Young, B.F. Clarke, The value of cardiovascular autonomic function tests: 10 years experience in diabetes. Diabetes Care **8**(5), 491–498 (1985)

82. M.H. Imam, C.K. Karmakar, A.H. Khandoker, H.F. Jelinek, M. Palaniswami, Heart rate independent QT variability component can detect subclinical cardiac autonomic neuropathy in diabetes, in *2016 IEEE 38th Annual International Conference of the Engineering in Medicine and Biology Society (EMBC)* (IEEE, New York, 2016), pp. 928–931

83. H.F. Jelinek, D.J. Cornforth, M.P. Tarvainen, N.T. Miloevic, Multiscale Renyi entropy and cardiac autonomic neuropathy, in *2015 20th International Conference on Control Systems and Computer Science (CSCS)* (IEEE, New York, 2015), pp. 545–547

84. H.F. Jelinek, M.P. Tarvainen, D.J. Cornforth, Renyi entropy in identification of cardiac autonomic neuropathy in diabetes, in *Computing in Cardiology (CinC), 2012* (IEEE, New York, 2012)

85. D.J. Cornforth, M.P. Tarvainen, H.F. Jelinek, Evaluation of normalised Renyi entropy for classification of cardiac autonomic neuropathy, in *2014 8th Conference of the European Study Group on Cardiovascular Oscillations (ESGCO)* (IEEE, New York, 2014), pp. 1–2

86. M.P. Tarvainen, D.J. Cornforth, H.F. Jelinek, Principal component analysis of heart rate variability data in assessing cardiac autonomic neuropathy, in *2014 36th Annual International Conference of the IEEE Engineering in Medicine and Biology Society (EMBC)* (IEEE, New York, 2014), pp. 6667–6670

87. J. Abawajy, A. Kelarev, M.U. Chowdhury, H.F. Jelinek, Enhancing predictive accuracy of cardiac autonomic neuropathy using blood biochemistry features and iterative multi tier ensembles. IEEE J. Biomed. Health Inf. **20**(1), 408–415 (2016)

88. P. Laguna, D. Vigo, R. Jane, P. Caminal, Automatic wave onset and offset determination in ECG signals: validation with the CSE database, in *Proceedings of Computers in Cardiology 1992* (IEEE, New York, 1992), pp. 167–170

89. J.M. Bote, J. Recas, F. Rincon, D. Atienza, R. Hermida, A modular low-complexity ECG delineation algorithm for real-time embedded systems. IEEE J. Biomed. Health Inf. **22**(2), 429–441 (2017)

90. T. Tekeste, H. Saleh, B. Mohammad, A. Khandoker, M. Elnaggar, A nano-Watt ECG feature extraction engine in 65nm technology. IEEE Trans. Circuits Syst. Express Briefs PP(99):1–1 (2017)

91. S.Y. Hsu, Y. Ho, Y. Tseng, T.Y. Lin, P.Y. Chang, J.W. Lee, J.H. Hsiao, S.M. Chuang, T.Z. Yang, P.C. Liu, T.F. Yang, R.J. Chen, C. Su, C.Y. Lee, A sub-100 μW multi-functional cardiac signal processor for mobile healthcare applications, in *2012 Symposium on VLSI Circuits (VLSIC)*, June 2012, pp. 156–157

92. M.D. Rollins, J.G. Jenkins, D.J. Carson, B.G. McClure, R.H. Mitchell, S.Z. Imam, Power spectral analysis of the electrocardiogram in diabetic children. Diabetologia **35**(5), 452–455 (1992)

93. D.J. Ewing, B.F. Clarke, Diagnosis and management of diabetic autonomic neuropathy. Br. Med. J. (Clin. Res. Ed.) **285**(6346), 916 (1982)

94. V. Spallone, G. Menzinger, Diagnosis of cardiovascular autonomic neuropathy in diabetes. Diabetes **46**(Suppl. 2), S67–S76 (1997)

95. M. Malik, Task force of the European Society of Cardiology and the North American Society of Pacing and Electrophysiology. Heart rate variability. Standards of measurement, physiological interpretation, and clinical use. Eur. Heart J. **17**, 354–381 (1996)

96. A.H. Khandoker, H.F. Jelinek, T. Moritani, M. Palaniswami, Association of cardiac autonomic neuropathy with alteration of sympatho-vagal balance through heart rate variability analysis. Med. Eng. Phys. **32**(2), 161–167 (2010)

97. E.K. Kasper, R.D. Bergeret et al., Beat to beat QT interval variability: novel evidence for repolarization lability in Ischemic and nonischemic dilated cardiomyopathy. Circulation **96**(5), 1557–1565 (1997)

98. A. L. Goldberger, L. A. N. Amaral, L. Glass, J. M. Hausdorff, P. C. Ivanov, R. G. Mark, J. E. Mietus, G. B. Moody, C.-K.Peng, and H. E.Stanley, "PhysioBank, PhysioToolkit, and PhysioNet: Components of a New Research Resource for Complex Physiologic Signals," Circulation, vol. 101, no. 23, pp. e215–e220, 2000 (June 13), circulation Electronic Pages: http://circ.ahajournals.org/cgi/content/full/101/23/e215PMID: 1085218; doi:10.1161/01.CIR.101.23.e215.

99. H.F. Jelinek, C. Wilding, P. Tinely, An innovative multi-disciplinary diabetes complications screening program in a rural community: a description and preliminary results of the screening. Aust. J. Prim. Health **12**(1), 14–20 (2006)

100. M. Yasin, T. Tekeste, H. Saleh, B. Mohammad, H. Saleh, O. Sinanoglu, M. Ismail, Ultra-low power, secure IoT platform for predicting cardiovascular diseases. IEEE Trans. Circuits Syst. Regul. Pap. **64**(9), 2624–2637 (2017)

101. H. Kim, R.F. Yazicioglu, T. Torfs, P. Merken, H.-J. Yoo, C. Van Hoof, A low power ECG signal processor for ambulatory arrhythmia monitoring system, in *Proceedings of IEEE Symposium on VLSI Circuits* (2010), pp. 19–20

Index

© Springer International Publishing AG, part of Springer Nature 2019
T. Tekeste Habte et al., *Ultra Low Power ECG Processing System
for IoT Devices*, Analog Circuits and Signal Processing,
https://doi.org/10.1007/978-3-319-97016-5

Printed in the United States
By Bookmasters